HOLT SCIENCE & TECHNOLOGY

Physical Science

DIRECTED READING WORKSHEETS

This book was printed with soy-based ink on acid-free recycled content paper, containing 10% POSTCONSUMER WASTE.

HOLT, RINEHART AND WINSTON

A Harcourt Classroom Education Company

Austin • New York • Orlando • Atlanta • San Francisco • Boston • Dallas • Toronto • London

Welcome!

Imagine that you have just entered a foreign land and encountered a unique culture. What better way to experience the unfamiliar territory than to find a knowledgeable guide? He or she could point out beautiful landscapes and historical landmarks while dazzling you with interesting tidbits about the region. Your guide could help you make the most of your visit and help make it a visit you'll remember.

Well you have just entered a foreign realm! You've entered the world of *Holt Science & Technology Physical Science*. To help you make the most of your journey, use this booklet as your personal guide. Your guide will help you focus your attention on interesting images and important scientific facts. Your guide will also offer tips to help you understand the local language and ask you questions along the way to make sure you don't miss anything.

So sit back and get ready to fully experience *Holt Science & Technology!* Don't worry, this guide knows the ropes—all you have to do is follow along!

Art and Photo Credits
All work, unless otherwise noted, contributed by Holt, Rinehart and Winston.
Abbreviated as follows: (t) top; (b) bottom; (l) left; (r) right; (c) center; (bkgd) background.
Front cover (owl), Kim Taylor/Bruce Coleman, Inc.; (bridge), Henry K. Kaiser/Leo de Wys; (dove), Stephen Dalton/Photo Researchers, Inc.

Printed in the United States of America

ISBN 0-03-054397-5   5 6 085 04 03 02

▪ CONTENTS ▪

CHAPTER

1 **DIRECTED READING WORKSHEET**

The World of Physical Science

As you read Chapter 1, which begins on page 4 of your textbook, answer the following questions.

Would You Believe . . . ? (p. 4)

1. How did James Czarnowski get his idea for the penguin boat, *Proteus*? Explain.

_____

_____

_____

_____

_____

2. What is unusual about the way that *Proteus* moves through the water?

_____

_____

_____

What Do You Think? (p. 5)

Answer these questions in your ScienceLog now. Then later, you'll have a chance to revise your answers based on what you've learned.

Investigate! (p. 5)

3. What is the seemingly impossible problem in this activity?

_____

_____

_____

Section 1: Exploring Physical Science (p. 6)

4. Your reflection on the inside of a spoon is different from your

reflection on the outside of a spoon. True or False? (Circle one.)

That's Science! (p. 6)

5. Which of the following activities are part of "doing science"? (Circle all that apply.)

   a. asking questions   **c.** making observations

   b. being curious   **d.** taking an opinion poll

6. Only scientists can do science. True or False? (Circle one.)

Matter + Energy → Physical Science (p. 7)

7. Physical science is the study of energy and the stuff that every-thing is made of. True or False? (Circle one.)

8. What do air, a ball, and a cheetah have in common?

_____

_____

9. All matter has energy, even if it isn't moving. True or False? (Circle one.)

10. What is one question you will answer as you explore physical science?

_____

_____

11. Chemistry and physics are both fields of

_____ . Chemists study the different forms

of _____ and how they interact.

_____ and how it affects

_____ are studied in physics.

Looking at the charts on page 8, identify the field of physical science to which each of the following descriptions belongs by writing *physics* or *chemistry* in the space provided.

12. _____ how a compass works

13. _____ why water boils at 100°C

14. _____ how chlorine and sodium combine to form table salt

15. _____ why you move to the right when the car you are in turns left

16. _____ why you see a rainbow after a rainstorm

Physical Science Is All Around You (p. 9)

Choose the term in Column B that best matches the topic of physical science in Column A, and write the corresponding letter in the space provided.

Column A	Column B
_____ **17.** waves, currents, and ocean water chemistry	**a.** ecology
_____ **18.** the transfer of energy in a food chain	**b.** botany
_____ **19.** weather patterns	**c.** oceanography
_____ **20.** how plants use carbon dioxide and water to make food	**d.** biology
_____ **21.** earthquake waves	**e.** astronomy
_____ **22.** the motion of galaxies in the universe	**f.** geology
_____ **23.** how the brain sends electrical impulses throughout the body	**g.** meteorology

Physical Science in Action (p. 10)

24. Which of the careers on page 10 sounds most interesting to you, and why?

_____

_____

_____

Review (p. 10)

Now that you've finished Section 1, review what you learned by answering the Review questions in your ScienceLog.

Section 2: Using the Scientific Method (p. 11)

1. Advancements in science are mostly due to luck.

True or False? (Circle one.)

What Is the Scientific Method? (p. 11)

2. Scientists use the scientific method to

_____ questions and

_____ problems by following a series of

_____ .

Chapter 1, continued

Ask a Question (p. 12)

3. Scientists usually ask a question before they make any

observations. True or False? (Circle one.)

4. _____ is the measure of how much energy
is put out by a system compared to how much energy is supplied
to a system.

Form a Hypothesis (p. 14)

5. A hypothesis is
   a. not based upon observations.
   b. usually stated in an "if . . . then . . ." format.
   c. a possible answer to a question.
   d. best when it is untestable.

Review (p. 14)

Now that you've finished the first part of Section 2, review what you
learned by answering the Review questions in your ScienceLog.

Test the Hypothesis (p. 15)

6. Explain the difference between a control group and an experimental group in a controlled experiment.

_____

_____

_____

_____

7. Scientists collect information called _____ .

Analyze the Results (p. 16)

8. Analysis of results means _____ data into
tables or graphs and looking for the relationships within the data.

Draw Conclusions (p. 17)

9. Which of the following is NOT true when drawing conclusions?
   a. The results may support the hypothesis.
   b. The results may disprove the hypothesis.
   c. The results may neither support nor disprove the hypothesis.
   d. None of the above

Communicate Results (p. 17)

10. What are three ways that communicating results can help other scientists further scientific research?

_____

_____

_____

_____

_____

11. Which of the following statements is ALWAYS true for scientific investigations?

    a. Scientists never have a clear idea of the problem they are trying to solve.

    b. Scientists keep testing the same hypothesis.

    c. Scientists take accurate measurements and accurately record data.

    d. Scientists follow the steps of the scientific method in the same order.

Building Scientific Knowledge (p. 18)

12. An idea that is supported by many tests and experiments can

    become a _____ or a

    _____ .

13. Which of the following is NOT true of a scientific theory?

    a. It unifies hypotheses and observations that have been supported by testing.

    b. It can predict an observation you might make in the future.

    c. It can be changed or replaced.

    d. It is a simple guess.

Mark each of the following statements *True* or *False*.

14. _____ You could be arrested if you break a scientific law.

15. _____ Scientific laws are determined by committee.

16. _____ Laws tell you *why* something happens, not *what* happens.

17. _____ A scientific law is a summary of many experimental results and observations.

18. Scientifically speaking, why do you think Figure 13 illustrates the big bang as a theory, not as a law?

_____

_____

_____

_____

_____

_____

_____

_____

_____

Review (p. 19)

Now that you've finished Section 2, review what you learned by answering the Review questions in your ScienceLog.

Section 3: Using Models in Physical Science (p. 20)

1. What did the MIT engineers hope to gain from making a model?

_____

_____

_____

What Is a Model? (p. 20)

2. You can represent an _____ or

_____ by using a model.

3. Which of the following are ways to use models in science? (Circle all that apply.)

a. looking at the tiny parts of a microscopic cell on a cell diagram

b. launching a homemade rocket in your backyard

c. observing how the parts of matter fit together without being able to see the tiny particles

d. testing a new design for a building on a computer before spending money on construction

4. How do you think a model rocket might help you understand a real rocket?

_____

_____

_____

_____

Models Help You Visualize Information (p. 21)

5. In Figure 15, how could using the spring toy as a model help you understand the behavior of sound waves?

_____

_____

_____

_____

6. When we picture things in our minds, we are creating models.

True or False? (Circle one.)

Models Are Just the Right Size (p. 22)

Decide whether a useful model for each of the following would be *larger* or *smaller* than the actual object, and write the appropriate answer in the space provided.

7. Mount Everest _____

8. a skyscraper _____

9. a computer chip _____

10. the moon _____

Models Build Scientific Knowledge (p. 22)

11. Models can be used as _____ to illustrate

_____ and _____

investigations.

12. Place each of the following statements in the correct sequence to explain how engineers could use *Proteus* to develop a new technology. Write the appropriate number in the space provided.

_____ Build a full-sized penguin boat.

_____ Discover what factors affect the model's efficiency.

_____ Conduct tests on the model.

13. Why do you think the model in Figure 19 might be useful for understanding atomic theory?

_____

_____

_____

_____

Models Can Save Time and Money (p. 23)

14. How can cyber-crashes like the one in Figure 20 save time and money?

_____

_____

_____

_____

Review (p. 23)

Now that you've finished Section 3, review what you learned by answering the Review questions in your ScienceLog.

Section 4: Measurement and Safety in Physical Science (p. 24)

1. At one time, systems of measurement were based on objects that varied in size, such as body parts and grains of barley.

   True or False? (Circle one.)

The International System of Units (p. 24)

2. Scientists can _____ and

   _____ their data when data are expressed in SI units.

Chapter 1, continued

3. Which unit of measurement is most practical for measuring a house?

    a. centimeters (cm)   **c.** meters (m)

    b. kilometers (km)   **d.** millimeters (mm)

4. Before you can determine how many lenses will fit into a moving crate, what information do you need?

_____

_____

_____

5. Which of the following is NOT a valid unit of measurement for volume?

    a. milliliter   **c.** microgram

    b. cubic centimeter   **d.** liter

6. Would you measure the mass of a car in milligrams? Why or why not? If not, which unit would you use?

_____

_____

_____

_____

_____

7. The _____ is the SI unit for temperature, but scientists often use degrees Celsius.

Derived Quantities (p. 26)

8. List two kinds of derived quantities.

_____

_____

9. What formula would you use to find out how much carpet it would take to cover the floor of your classroom? Write the equation.

_____

_____

10. How could you calculate the mass per unit of volume of the gear in the right column of page 27?

_____

_____

_____

_____

Safety Rules! (p. 27)

Match the safety icon in Column B to the correct description in Column A, and write the corresponding letter in the space provided.

Column A	Column B
____ **11.** hand safety	**a.**
____ **12.** sharp object	**b.**
____ **13.** clothing protection	**c.**
____ **14.** chemical safety	**d.**
____ **15.** eye protection	**e.**
____ **16.** electrical safety	**f.**
____ **17.** plant safety	**g.**
____ **18.** heating safety	**h.**
____ **19.** animal safety	**i.**

Review (p. 27)

Now that you've finished Section 4, review what you learned by answering the Review questions in your ScienceLog.

CHAPTER

2 DIRECTED READING WORKSHEET

The Properties of Matter

As you read Chapter 2, which begins on page 34 of your textbook, answer the following questions.

Imagine . . . (p. 34)

1. What are two ways to tell "fool's gold" from real gold?

_____

_____

_____

_____

_____

_____

_____

What Do You Think? (p. 35)

Answer these questions in your ScienceLog now. Then later, you'll have a chance to revise your answers based on what you've learned.

Investigate! (p. 35)

2. What will you do in this activity?

_____

_____

_____

Section 1: What Is Matter? (p. 36)

3. What do a human, hot soup, and a neon sign have in common?

_____

_____

Everything Is Made of Matter (p. 36)

4. Anything that has _____ and

_____ is called matter.

_____

Chapter 2, continued

_____

Matter Has Volume (p. 36)

Mark each of the following statements *True* or *False*.

5. _____ An object's volume is the amount of space the object takes up.

6. _____ Things with volume can't share the same space at the same time.

7. _____ When you measure a volume of water in a graduated cylinder, you should look at the bottom of the meniscus.

8. _____ A liquid's volume is usually expressed in grams or milligrams.

9. The volume of solid objects is expressed in

_____ units. One milliliter is equal to

_____ .

10. What three dimensions do you need to find the volume of a rectangular solid object?

_____

_____

11. You can't use a ruler to measure a gas, and you can't pour it into a graduated cylinder. So how do you find its volume?

_____

_____

_____

Matter Has Mass (p. 38)

12. List the following objects in order from the least mass to the greatest mass: an elephant, a hamster, a skyscraper, the moon.

_____

13. The mass of an object should be constant, yet the mass of the puppy in Figure 5, on page 39, will change over time. Explain.

_____

_____

_____

_____

Chapter 2, continued

The Difference Between Mass and Weight (p. 39)

14. Why does all matter experience gravity?

    a. All matter has volume.   **c.** All matter is constant.

    b. All matter has mass.   **d.** All matter is stable.

15. Look at Figure 6 on page 39. As two objects get

    _____ , the force of gravity between them

    increases. (closer together or farther apart)

16. Weight is a measure of the gravitational force exerted on an

    object. True or False? (Circle one.)

17. A brick weighs less in space than it does on Earth. Why?

    _____

    _____

    _____

    _____

18. Why do people tend to confuse the terms *mass* and *weight*?
(Circle all that apply.)

    a. Both remain constant on Earth.

    b. People use the terms interchangeably.

    c. Mass is the same thing as weight.

    d. Mass is also dependent on gravity.

Measuring Mass and Weight (p. 41)

19. The unit for _____ is the kilogram. The

    unit for _____ is the newton.

Mass Is a Measure of Inertia (p. 42)

20. How do mass and inertia make it easier to pick up an empty
juice bottle than a full juice bottle?

    _____

    _____

    _____

    _____

Review (p. 42)

Now that you've finished Section 1, review what you learned by
answering the Review questions in your ScienceLog.

CHAPTER 2

_____

Section 2: Describing Matter (p. 43)

1. In a game of 20 Questions, the most helpful questions you can

ask are about the _____ of the object.

Physical Properties (p. 43)

2. Which physical properties is the person in the picture on page 43
asking about?

_____

_____

_____

Match each physical property in Column B to the correct phrase in
Column A, and write the corresponding letter in the appropriate
space. Use the table on page 44 to help you.

Column A	Column B
_____ **3.** Sand does not dissolve in water.	**a.** state
_____ **4.** Gold can be made into gold foil.	**b.** thermal conductivity
_____ **5.** Ice is the solid form of water.	**c.** solubility
_____ **6.** Copper can be drawn out into wire.	**d.** density
_____ **7.** A foam cup protects your hand from being burned by the hot chocolate the cup contains.	**e.** ductility
_____ **8.** Ice cubes float in a glass of water.	**f.** malleability

9. In the formula for density, D means _____ ,

V stands for _____ , and m stands for

_____ .

10. What are two reasons why density is a useful property for
identifying substances?

_____

_____

_____

_____

_____

Chapter 2, continued

11. Using the table on page 45, list the elements mercury, water, gold, and oxygen from the densest to the least dense.

_____

12. What will happen if you shake the jar in Figure 12? Explain.

_____

_____

_____

_____

_____

_____

_____

13. Is 40 mL of oil denser than 25 mL of vinegar? _____

14. Density is dependent on the amount of substance you have.

True or False? (Circle one.)

Review (p. 46)

Now that you've finished the first part of Section 2, review what you've learned by answering the Review questions in your ScienceLog.

Chemical Properties (p. 47)

15. Which of the following are true of chemical properties? (Circle all that apply.)

    a. They indicate a substance's ability to change identity.
    b. They indicate one substance's ability to react with another.
    c. They describe matter.
    d. They can be observed with your senses.

16. In Figure 13, on page 47, why isn't there rust in the painted areas of the car?

_____

_____

_____

_____

Chapter 2, continued

Physical vs. Chemical Properties (p. 48)

17. Characteristic properties help scientists to distinguish one

substance from another. True or False? (Circle one.)

18. Which of the following represents ONLY physical properties? The
table on page 48 may help you.

   a. flammable, dense, malleable
   b. malleable, reactive, dense
   c. powdery, dense, red
   d. clear, grainy, nonflammable

Physical Changes Don't Form New Substances (p. 48)

19. What is a physical change?

_____

_____

20. If you make a physical change to a substance, the identity of the

substance changes. True or False? (Circle one.)

21. Many physical changes are _____ to
undo. (easy or difficult)

Chemical Changes Form New Substances (p. 49)

22. A chemical _____ describes a substance's

ability to change. A chemical _____
occurs when a substance turns into another substance.

23. How do you know that baking a cake involves chemical changes?

_____

_____

_____

24. In Examples of Chemical Changes, the odor of sour milk indicates

a chemical change has taken place. True or False? (Circle one.)

25. Some chemical changes can be reversed with more chemical

changes. True or False? (Circle one.)

Review (p. 51)

Now that you've finished Section 2, review what you learned by
answering the Review questions in your ScienceLog.

CHAPTER

3   **DIRECTED READING WORKSHEET**

States of Matter

As you read Chapter 3, which begins on page 58 of your textbook, answer the following questions.

Imagine . . . (p. 58)

1. Lightning sometimes leaves behind a strange calling card. What is it called and how is it formed?

   _____

   _____

   _____

   _____

2. How do glassmakers use a change of state to make light bulbs, windows, and bottles?

   _____

   _____

   _____

What Do You Think? (p. 59)

Answer these questions in your ScienceLog now. Then later, you'll have a chance to revise your answers based on what you've learned.

Investigate! (p. 59)

3. The purpose of this activity is
   a. to disinfect your hand with alcohol.
   b. to observe a change of state of alcohol.
   c. to observe a change of state of water.

Section 1: Four States of Matter (p. 60)

4. Look at Figure 1. Which of the following states of matter does Hero's steam engine demonstrate? (Circle all that apply.)
   a. solid   **c.** gas
   b. liquid   **d.** plasma

Moving Particles Make Up All Matter (p. 60)

5. The speed of the particles and the strength of the attraction

   between them determine the _____ of the

   substance.

Match the state of matter in Column B with the description in
Column A, and write the corresponding letter in the appropriate space.

Column A	Column B
_____ **6.** Particles have a strong attraction to each other.	**a.** solid
_____ **7.** Particles move independently of each other.	**b.** liquid
_____ **8.** Particles are able to slide past one another but do not move independently of each other.	**c.** gas
_____ **9.** Particles vibrate in place.	
_____ **10.** Particles move fast enough to overcome nearly all of the attraction between them.	

Solids Have Definite Shape and Volume (p. 61)

11. The ship in the bottle in Figure 3 is a solid. How can you tell?

_____

_____

_____

12. Particles that are arranged in a repeating pattern of rows form

amorphous solids. True or False? (Circle one.)

Liquids Change Shape but Not Volume (p. 62)

13. How do the particles of a liquid make it possible to pour juice
into a glass?

_____

_____

_____

14. What does Figure 6 show you about the properties of a liquid?

_____

_____

15. Liquids tend to form in spherical droplets because of

_____ tension.

16. Water has a lower _____ than honey.

Chapter 3, continued

Gases Change Both Shape and Volume (p. 63)

17. How is it possible for a cylinder of helium to fill 700 balloons?

_____

_____

_____

_____

_____

Gas Under Pressure (p. 64)

18. The amount of _____ exerted on a given area is called pressure.

Review (p. 64)

Now that you've finished the first part of Section 1, review what you learned by answering the Review questions in your ScienceLog.

Laws Describe Gas Behavior (p. 65)

19. The volume of a gas is always the volume of its container.

True or False? (Circle one.)

20. Boyle's law states that if you keep the temperature constant for a fixed amount of gas, a decrease in pressure means a(n)

_____ in the volume of the gas.

21. Weather balloons are only partially inflated before they're released into the atmosphere. Why is that?

_____

_____

_____

_____

22. _____ is demonstrated by putting a balloon in the freezer.

23. All of the following remain constant in Figure 11 EXCEPT

　　a. the type of piston. 　　**c.** the volume of the gas.
　　b. the amount of gas. 　　**d.** the pressure.

Chapter 3, continued

Plasmas (p. 67)

Mark each of the following statements *True* or *False*.

24. _____ More than 99 percent of the known matter in the universe is in the plasma state.

25. _____ Plasmas are made up of particles that have broken apart.

26. _____ Plasmas have a definite shape and volume.

27. _____ Plasmas and gases conduct electric current.

28. _____ Plasmas are affected by magnetic fields.

29. Lightning and fire are examples of _____ plasmas.

30. The incredible light show in Figure 12, on page 67, is caused by plasma. How?

_____

_____

_____

Review (p. 67)

Now that you've finished Section 1, review what you learned by answering the Review questions in your ScienceLog.

Section 2: Changes of State (p. 68)

1. When a substance changes from one _____ form to another, we say the substance has had a change of state.

2. List the five changes of state.

_____

Energy and Changes of State (p. 68)

3. The identity of a substance changes during a change of state.

   True or False? (Circle one.)

4. Temperature is the measure of the speed of particles.

   True or False? (Circle one.)

5. Temperature is a transfer of energy. True or False? (Circle one.)

6. Which has the most energy?

   a. particles in steam   **c.** particles in ice

   b. particles in liquid water   **d.** particles in freezing water

_____

Chapter 3, continued

Melting: Solids to Liquids (p. 69)

7. Could you use gallium to make jewelry? Why or why not?

_____

_____

_____

8. Melting point is a characteristic property, because it is the same
for different amounts of the same substance. True or False?
(Circle one.)

Freezing: Liquids to Solids (p. 69)

9. A substance's freezing point is the temperature at which it

changes from a _____ to a

_____ .

10. What happens if energy is added or removed from the ice water
in Figure 15?

_____

_____

_____

11. Freezing is considered an exothermic change because

_____ is removed from the substance.

Vaporization: Liquids to Gases (p. 70)

Choose the term in Column B that best matches the description in
Column A, and write the corresponding letter in the space provided.

Column A	Column B
____ **12.** vaporization at the surface of a liquid below its boiling point	**a.** boiling point
____ **13.** the change of state from a liquid to a gas	**b.** vaporization
____ **14.** vaporization that occurs throughout a liquid	**c.** steam
____ **15.** the product of vaporization of liquid water	**d.** evaporation
____ **16.** temperature at which a liquid boils	**e.** boiling

---

Chapter 3, continued

Condensation: Gases to Liquids (p. 71)

Mark each of the following statements *True* or *False*.

17. _____ At a given pressure, the condensation point for a substance is the same as its melting point.

18. _____ For a substance to change from a gas to a liquid, particles must clump together.

19. _____ Condensation is an exothermic change.

Sublimation: Solids Directly to Gases (p. 72)

20. Solid carbon dioxide isn't ice. So why is it called "dry ice"?

_____

_____

_____

21. The change of state from a solid to a _____ is called sublimation. Energy must be added for sublimation to occur, so it is an _____ change.

Comparing Changes of State (p. 72)

22. Look at the table on page 72. Which two changes of state occur at the same temperature?

    a. condensation and melting
    b. sublimation and freezing
    c. vaporization and condensation
    d. melting and vaporization

Temperature Change Versus Change of State (p. 73)

Mark each of the following statements *True* or *False*. Figure 19 may help you.

23. _____ The speed of the particles in a substance changes when the temperature changes.

24. _____ The temperature of a substance changes before the change of state is complete.

25. _____ Energy must be added to a substance to move its temperature from the melting point to the boiling point.

Review (p. 73)

Now that you've finished Section 2, review what you learned by answering the Review questions in your ScienceLog.

Name _____ Date _____ Class _____

Elements, Compounds, and Mixtures

As you read Chapter 4, which begins on page 80 of your textbook, answer the following questions.

This Really Happened! (p. 80)

1. How do scientists think that the composition of the *Titanic*'s hull caused the "unsinkable" ship to sink?

_____

_____

_____

_____

2. Why might it be important to learn about the properties of elements, mixtures, and compounds?

_____

_____

_____

What Do You Think? (p. 81)

Answer these questions in your ScienceLog now. Then later, you'll have a chance to revise your answers based on what you've learned.

Investigate! (p. 81)

3. What do you think will happen to the ink of the black marker in this activity?

_____

_____

_____

_____

CHAPTER 4

Section 1: Elements (p. 82)

4. What physical changes can you make to a substance to determine if it's an element? (Circle all that apply.)

  a. crushing   **c.** filtering
  b. grinding   **d.** passing electric current

An Element Has Only One Type of Particle (p. 82)

5. A pure substance is a substance that contains only one type of particle. True or False? (Circle one.)

6. In Figure 2, what do the skillet and the meteorite have in common?

_____

_____

Every Element Has a Unique Set of Properties (p. 83)

7. Characteristic properties of elements do NOT depend on the amount of material present in a sample of the element.

True or False? (Circle one.)

8. Why does a helium-filled balloon float up when it is released?

_____

_____

9. Look at the properties listed below. Circle the characteristic properties of elements.

  size   melting point   density   shape

  mass   volume   color   surface area

  hardness   flammability   weight   reactivity with acid

10. Suppose you have a cube of nickel and a cube of cobalt, but you don't know which is which. How could you use the characteristic properties listed in Figure 3 to figure out which cube is nickel and which is cobalt?

_____

_____

_____

_____

_____
Chapter 4, continued
_____

Elements Are Classified by Their Properties (p. 84)

11. What are some common properties that most terriers share?

_____

_____

_____

12. Which of the following is a property that nickel, iron, and cobalt DON'T share?

   a. shiny

   b. poor conductivity of electric current

   c. good conductivity of thermal energy

   d. None of the above

13. All elements can be classified as metals, metalloids, or

nonmetals. True or False? (Circle one.)

Look at the chart on page 85. Match the categories of elements in Column B with the correct properties in Column A, and write the corresponding letter in the appropriate space. Categories may be used more than once.

Column A	Column B
____ **14.** malleable	**a.** metalloids
____ **15.** dull or shiny	**b.** nonmetals
____ **16.** poor conductors	**c.** metals
____ **17.** tend to be brittle and unmalleable as solids	
____ **18.** always shiny	
____ **19.** also called semiconductors	
____ **20.** graphite in pencils	
____ **21.** always dull	
____ **22.** used in computer chips	
____ **23.** ductile	

Review (p. 85)

Now that you've finished Section 1, review what you learned by answering the Review questions in your ScienceLog.

Section 2: Compounds (p. 86)

1. When two or more elements are chemically combined to form a new pure substance, we call that new substance a

   _____ .

2. A compound is different from the elements that reacted to

   form it. True or False? (Circle one.)

3. List three examples of compounds you encounter every day.

   _____

Elements Combine in a Definite Ratio to Form a Compound (p. 86)

4. Which of the following is NOT true about compounds?
   a. Compounds join in specific ratios according to their masses.
   b. Mass ratios can be written as a ratio or a fraction.
   c. Compounds are random combinations of elements.
   d. Different mass ratios mean different compounds.

Every Compound Has a Unique Set of Properties (p. 87)

Mark each of the following statements *True* or *False*.

5. _____ Each compound has its own physical properties.

6. _____ Compounds cannot be identified by their chemical properties.

7. _____ A compound has the same properties as the elements that form it.

8. Sodium and chlorine can be extremely dangerous in their elemental form. So how is it possible that we can eat them in a compound?

   _____

   _____

   _____

   _____

   _____

   _____

   _____

Chapter 4, continued

Compounds Can Be Broken Down into Simpler Substances (p. 88)

9. What compound helps to give carbonated beverages their "fizz"?

_____

_____

Which elements make up this compound?

_____

_____

_____

10. A physical change is the only way to break down a compound.

True or False? (Circle one.)

11. Look at the Physics Connection. The chemical process used to obtain industrial products, such as hydrogen peroxide, is

called _____ .

Compounds in Your World (p. 89)

12. Which of the following methods are used by living organisms to obtain nitrogen, an element needed to make proteins? (Circle all that apply.)

a. Bacteria on the roots of pea plants make compounds from atmospheric hydrogen.

b. Plants use nitrogen compounds in the soil.

c. Animals digest plants or animals that have eaten plants.

d. Plants take in carbon dioxide to make sugar.

13. What do most fertilizers, food preservatives, and medicines have in common?

_____

_____

_____

14. The compound _____ is broken down to produce the element used in cans, airplanes, and building materials.

Review (p. 89)

Now that you've finished Section 2, review what you learned by answering the Review questions in your ScienceLog.

CHAPTER 4

Section 3: Mixtures (p. 90)

1. A pizza is not a mixture. True or False? (Circle one.)

Properties of Mixtures (p. 90)

2. When two or more materials combine without reacting with
each other, they form a mixture. True or False? (Circle one.)

3. How do the granite in Figure 11 and the pizza at the top of the
page show you that the identity of a substance doesn't change in
a mixture?

   _____

   _____

   _____

   _____

   _____

4. Mixtures are separated through _____
changes.

Look at the pictures on page 91. Match the technique for separating
a mixture in Column B with the substances in Column A, and write
the corresponding letter in the appropriate space.

Column A	Column B
____ **5.** crude oil	**a.** distill the mixture
____ **6.** aluminum and iron	**b.** centrifuge the mixture
____ **7.** parts of the blood	**c.** filter the mixture
____ **8.** sulfur and water	**d.** pass a magnet over the mixture

9. Granite can be pink or black, depending on the ratio of feldspar,
mica, and quartz. True or False? (Circle one.)

Review (p. 92)
Now that you've finished the first part of Section 3, review what you
learned by answering the Review questions in your ScienceLog.

Solutions (p. 92)

10. Which of the following is NOT true of solutions?

    a. They contain a dissolved substance called a solute.

    b. They are composed of two or more evenly distributed substances.

    c. They contain a substance called a solvent, in which another substance is dissolved.

    d. They appear to be more than one substance.

11. In a gaseous or liquid solution, the volume of solvent is

_____ the volume of solute.
(less than or greater than)

12. The solid solution used to build the ship *Titanic* was a(n)

_____ called steel.

13. Which of the following is true of particles in solutions?

    a. Particles scatter light.

    b. Particles settle out.

    c. Particles can't be filtered out of these mixtures.

    d. Particles are large.

14. Concentration is a measure of the volume of a solution.

True or False? (Circle one.)

15. What is the difference between a concentrated solution and a dilute solution?

_____

_____

16. When a solution is holding all the solute it can hold at a given

temperature, we say the solution is _____ .

17. Solubility is not dependent on temperature. True or False? (Circle one.)

18. Solubility is expressed in grams of solute per

_____ of solvent.

19. Solubility of *gases* in liquids tends to _____

with an increase in temperature. Solubility of *solids* in liquids

tends to _____ with an increase in temperature.

20. What are three ways to make a sugar cube dissolve more quickly in water?

_____

_____

Suspensions (p. 96)

21. Which of the following does NOT describe a suspension?

    a. Particles are soluble.
    b. Particles settle out over time.
    c. Particles can be seen.
    d. Particles scatter light.

22. Look at the Biology Connection at the top of the page. Why is blood a suspension?

_____

_____

Colloids (p. 97)

23. What do gelatin, milk, and stick deodorant have in common?

_____

Match the mixtures in Column B to the characteristics in Column A, and write the corresponding letter in the appropriate space. Mixtures may be used more than once.

Column A	Column B
_____ **24.** Particles do not settle out.	**a.** colloids and suspensions
_____ **25.** Particles are larger.	
_____ **26.** Particles scatter a beam of light.	**b.** colloids and solutions
_____ **27.** Particles cannot be filtered out.	

28. Look at Figure 18. How can a colloid be dangerous to drivers?

_____

_____

Review (p. 97)

Now that you've finished Section 3, review what you learned by answering the Review questions in your ScienceLog.

5 **DIRECTED READING WORKSHEET**

Matter in Motion

As you read Chapter 5, which begins on page 106 of your textbook, answer the following questions.

Would You Believe . . . ? (p. 106)

1. One reason Native Americans played the game of lacrosse was for fun. What was the other reason?

_____

_____

2. What are the advantages of using a lacrosse stick? (Circle all that apply.)

 a. It makes it possible to throw the ball at speeds over 100 km/h.
 b. It helps the defense players hold back the offense.
 c. It makes it possible to throw the ball over 100 km.
 d. It protects the hand from injury from the high speed ball.

What Do You Think? (p. 107)

Answer these questions in your ScienceLog now. Then later, you'll have a chance to revise your answers based on what you've learned.

Investigate! (p. 107)

3. What do you predict this activity will demonstrate?

_____

_____

_____

Section 1: Measuring Motion (p. 108)

4. Name something in motion that you can't see moving.

_____

_____

_____

Observing Motion (p. 108)

5. To determine if an object is in motion, compare its position over

 time to a _____ point.

_____

Chapter 5, continued

_____

6. Buildings, trees, and mountains are all useful reference points.
   Why?

   _____

   _____

7. Can a moving object be used as a reference point? Explain.

   _____

   _____

   _____

Speed Depends on Distance and Time (p. 109)

Each of the following statements is false. Change the underlined
word to make the statement true. Write the new word in the space
provided.

8. <u>Motion</u> is the rate at which an object moves.

   _____

9. How fast an object moves depends on the distance traveled and
   the <u>road</u> taken to travel that distance.

   _____

10. The SI unit for speed is <u>km/h</u>.

    _____

11. Why is it useful to calculate average speed?

    _____

    _____

12. Write out in words how to calculate average speed.

    _____

    _____

    _____

13. Look at the Brain Food on p. 109. Suppose a car travels 250 m in
    10 seconds. Is its average speed greater than or less than that of a
    running cheetah?

    _____

Chapter 5, continued

Velocity: Direction Matters (p. 110)

14. Why don't the birds in the riddle end up at the same destination?

_____

_____

15. Velocity has speed and _____ .

16. Which of the following does NOT experience a change in velocity?

    a. A motorcyclist driving down a straight street applies the brakes.

    b. While maintaining the same speed and direction, an experimental car switches from gasoline to electric power.

    c. A baseball player running from first base to second base at 10 m/s comes to a stop in 1.5 seconds.

    d. A bus traveling at a constant speed turns a corner.

17. To find the resultant velocity, add velocities that are in

_____ direction(s). Subtract velocities

that are in _____ direction(s).

Review (p. 111)

Now that you've finished the first part of Section 1, review what you learned by answering the Review questions in your ScienceLog.

Acceleration: The Rate at Which Velocity Changes (p. 112)

18. Why did the neighbor say you had great acceleration as you slowed down and swerved to avoid hitting a rock?

_____

_____

_____

_____

_____

19. Write the formula for calculating acceleration in the space below.

CHAPTER 5

20. Scientifically speaking, how do you know the cyclist in Figure 4, on page 113, is accelerating?

_____

_____

_____

_____

_____

_____

21. Another name for acceleration in which velocity increases is

_____ acceleration.

22. Negative acceleration, or acceleration in which velocity

decreases, is also called _____ .

23. What kind of acceleration is occurring in Figure 5, on page 114?

_____

24. When you are standing completely still at the equator, you are

accelerating. True or False? (Circle one.)

25. How can you recognize acceleration on a graph?

    a. The graph shows distance versus time.
    b. The graph shows time versus distance.
    c. The graph shows velocity changing as time passes.
    d. The graph is a straight line.

Review (p. 114)

Now that you've finished Section 1, review what you learned by answering the Review questions in your ScienceLog.

Section 2: What Is a Force? (p. 115)

Mark the following statements *True* or *False*.

1. _____ All forces have size and direction.

2. _____ A force is a push or a pull.

3. _____ Forces are expressed in liters.

Chapter 5, continued

Forces Act on Objects (p. 115)

4. You can exert a push without there being an object to receive the

push. True or False? (Circle one.)

5. Name three examples of objects that you exert forces on when
you are doing your schoolwork.

_____

_____

_____

_____

_____

6. In which of the following situations is a force being exerted?
(Circle all that apply.)

   a. A woman pushes the elevator button.
   b. A pile of soil sits on the ground.
   c. Socks like the ones in Figure 7, on page 116, cling together
      when they have just come out of the dryer.
   d. Magnets stick to the refrigerator.

Forces in Combination (p. 116)

7. In Figure 8, on page 116, how does the net force help the students
move the piano?

_____

_____

_____

_____

8. Suppose the dog on the left in Figure 9, on page 117, increased its
force to 13 N. Which dog would win the tug-of-war? Explain.

_____

_____

_____

_____

Chapter 5, continued

Unbalanced and Balanced Forces (p. 117)

9. Why is it useful to know the net force?

_____

_____

10. Forces are unbalanced when the net force is not equal to

_____ .

11. To start or change the motion of an object, you need a(n)

_____ force. (balanced or unbalanced)

12. Forces are balanced when the net force applied to an object is

_____ zero.
(less than, greater than, or equal to)

13. Are the forces on the cards in Figure 10 balanced? How do you know?

_____

_____

Review (p. 118)

Now that you've finished Section 2, review what you learned by answering the Review questions in your ScienceLog.

Section 3: Friction: A Force that Opposes Motion (p. 119)

1. What force is responsible for the painful difference between sliding on grass and sliding on pavement?

_____

_____

The Source of Friction (p. 119)

2. Friction occurs when the hills and valleys of two surfaces stick

together. True or False? (Circle one.)

3. Pavement creates more friction than grass. Why is that?

_____

_____

Chapter 5, continued

---

4. Why is more force needed to slide the larger book in Figure 12, on page 120?

_____

_____

_____

_____

5. Friction is affected by the amount of surface that is touching.

True or False? (Circle one.)

Types of Friction (p. 121)

Match each type of friction in Column B with its example in Column A, and write the corresponding letter in the space provided.

Column A	Column B
____ **6.** a hockey puck crossing an ice rink	**a.** sliding friction
____ **7.** a crate resting on a loading ramp	**b.** rolling friction
____ **8.** wheeled cart being pushed	**c.** fluid friction
____ **9.** air rushing past a speeding car	**d.** static friction

10. Static friction is at work if you try to drag a heavy suitcase along

the floor and the suitcase _____ .
(moves or doesn't move)

11. As soon as an object starts moving, static friction

_____ . (increases or disappears)

Friction Can Be Harmful or Helpful (p. 123)

12. How does friction harm the engine of a car?

_____

_____

_____

13. Why do you need friction to walk?

_____

_____

14. Which of the following are ways to reduce friction?
(Circle all that apply.)

   a. Use a lubricant.

   b. Make rubbing surfaces smoother.

   c. Push surfaces together.

   d. Change sliding friction to rolling friction.

Review (p. 124)

Now that you've finished Section 3, review what you learned by
answering the Review questions in your ScienceLog.

Section 4: Gravity: A Force of Attraction (p. 125)

1. Why did the astronauts in Figure 18 bounce on the moon?

_____

_____

2. The force of attraction between two objects due to their masses

is the force of _____ .

All Matter Is Affected by Gravity (p. 125)

3. Does all matter experience gravity? Explain.

_____

_____

_____

4. The force that pulls you toward your pencil is called the

_____ force.

5. Look at the Biology Connection at the bottom left of page 125.

Scientists think seeds can sense gravity. True or False? (Circle one.)

6. Since all objects are attracted to each other due to gravity, why
can't you see the objects moving toward each other?

_____

_____

_____

_____

Chapter 5, continued

---

7. How are objects around us affected by the mass of the Earth?

_____

_____

_____

_____

_____

The Law of Universal Gravitation (p. 126)

8. What did Newton figure out about the moon and a falling apple?

_____

_____

_____

_____

9. Newton's law of universal gravitation describes the relationships between all of the following EXCEPT

    a. distance.   **c.** heat.

    b. mass.   **d.** gravitational force.

10. Which of the following objects are subject to the law of universal gravitation? (Circle all that apply.)

    a. satellites   **c.** frogs

    b. water   **d.** stars

11. If the distance between the objects are the same, the gravitational force between two feathers is greater than the gravitational

force between two bowling balls. True or False? (Circle one.)

12. If two objects are moved _____ each other, the gravitational force between them increases. (away from or toward)

13. Why is a cat easier to pick up than an elephant?

_____

_____

14. Read the Astronomy Connection at the bottom right of page 127. In a _____ , gravity is so great that even light can't escape.

▲▲ CHAPTER 5 ▲

15. Why doesn't the sun's gravitational force pull you off the Earth?

_____

_____

_____

16. What would happen to the Earth and other planets in the solar system without the sun's gravitational force?

_____

_____

_____

Weight Is a Measure of Gravitational Force (p. 128)

17. The gravitational force exerted by an object depends on the

_____ of the object. The measure of the Earth's gravitational force on an object is the object's

_____.

Identify each of the following statements as describing mass or weight. Write *M* for mass and *W* for weight.

18. _____ different on the moon

19. _____ expressed in newtons

20. _____ expressed in grams

21. _____ measure of gravitational force

22. _____ value doesn't change

23. _____ amount of matter in an object

24. On Earth, mass and weight are constant, which means they are the same thing. True or False? (Circle one.)

Review (p. 129)

Now that you've finished Section 4, review what you learned by answering the Review questions in your ScienceLog.

CHAPTER

6 **DIRECTED READING WORKSHEET**

Forces in Motion

As you read Chapter 6, which begins on page 136 of your textbook, answer the following questions.

Imagine . . . (p. 136)

1. What is the Vomit Comet?

_____

_____

_____

_____

2. In this chapter you will learn how _____

affects the _____ of objects and how the

_____ of _____

apply to your life.

What Do You Think? (p. 137)

Answer these questions in your ScienceLog now. Then later, you'll have a chance to revise your answers based on what you've learned.

Investigate! (p. 137)

3. What is the purpose of this activity?

_____

_____

_____

Section 1: Gravity and Motion (p. 138)

4. Do you agree with what Aristotle might say, that the baseball would land first, then the marble? Explain.

_____

_____

_____

_____

Chapter 6, continued

All Objects Fall with the Same Acceleration (p. 138)

5. Did Galileo prove Aristotle wrong? Explain.

_____

_____

_____

_____

_____

6. What does 9.8 m/s/s have to do with acceleration?

_____

_____

_____

Air Resistance Slows Down Acceleration (p. 139)

7. Why does a crumpled piece of paper hit the ground before a flat sheet of paper?

_____

_____

8. Air resistance is affected by the _____

and _____ of an object.

9. Air resistance matches the _____ when the net force equals zero. (acceleration or force of gravity)

10. When a falling object stops _____ , it has

reached _____ velocity.

11. If there were no air resistance, hailstones would
   a. hit the Earth at velocities near 350 m/s.
   b. float gently to the ground like snowflakes.
   c. melt before they hit the ground.
   d. behave exactly as they do now.

12. A sky diver experiences free fall. True or False? (Circle one.)

13. Free fall occurs because of high air resistance. True or False? (Circle one.)

Orbiting Objects Are in Free Fall (p. 141)

14. An astronaut is weightless in space. True or False? (Circle one.)

15. The shuttle in Figure 7, on page 142, follows the curve of the Earth's surface as it moves _____ at a constant speed. At the same time, it is in _____ because of the Earth's gravity.

16. Why don't astronauts hit their head on the ceiling of the falling shuttle?

_____

_____

_____

_____

17. Earth's gravity provides a _____ force that keeps the moon in orbit.

Projectile Motion and Gravity (p. 143)

18. The projectile motion of a leaping frog has two components— _____ and _____ .

Mark each of the following statements *True* or *False*.

19. _____ The components of projectile motion affect each other.

20. _____ Horizontal motion of an object is parallel to the ground.

21. _____ Ignoring air resistance, the horizontal velocity of a thrown object never changes.

22. _____ On Earth, gravity gives thrown objects their downward vertical motion.

23. If you shoot an arrow aimed directly at the bull's-eye of your target, where will the arrow hit your target? Why?

_____

_____

_____

Review (p. 144)

Now that you've finished Section 1, review what you've learned by answering the Review questions in your ScienceLog.

_____

Chapter 6, continued

_____

Section 2: Newton's Laws of Motion (p. 145)

1. In 1686, _____ published *Principia,*

a work explaining _____ laws to
help people understand how forces relate to the

_____ of objects.

Newton's First Law of Motion (p. 145)

2. What is Newton's first law?

_____

_____

_____

_____

_____

3. An object in motion would keep moving forever if it never ran

into another object or an unbalanced force. True or False?
(Circle one.)

4. _____ is the unbalanced force that slows
down sliding desks, rolling baseballs, and moving cars.

5. How does inertia explain why it would be so difficult to play
softball with a bowling ball?

_____

_____

_____

_____

_____

_____

_____

Newton's Second Law of Motion (p. 148)

6. What is Newton's second law of motion?

_____

_____

_____

_____

7. Look at the Environment Connection in the upper right of page 148. A small car with a small engine cannot accelerate as well as a large car with a large engine. True or False? (Circle one.)

8. An object's acceleration decreases as the force on it increases. True or False? (Circle one.)

9. Force equals _____ times

_____ .

10. The watermelon in Figure 16, on page 149, has more

_____ and _____ than the apple, so the watermelon is harder to move than the apple.

Review (p. 149)

Now that you've finished the first part of Section 2, review what you've learned by answering the Review questions in your ScienceLog.

Newton's Third Law of Motion (p. 150)

11. What is Newton's third law of motion?

_____

_____

_____

_____

12. The phrase "equal and opposite" means that the action force and the reaction force have the same _____

but act in opposite _____ .

13. What action and reaction forces are present when you are sitting on a chair?

_____

_____

_____

_____

_____

14. In a force pair, the reaction and action forces affect the same object. True or False? (Circle one.)

15. When a ball falls off a ledge, gravity pulls the ball toward Earth and also pulls Earth toward the ball. True or False? (Circle one.)

Momentum Is a Property of Moving Objects (p. 152)

16. Why does it take longer for a large truck to stop than it does for a compact car to stop, even though they are traveling at the same velocity and the same braking force is applied?

_____

_____

17. Momentum depends on the _____ and

_____ of an object.

18. In Figure 19, on page 152, during the collision, the momentum of the cue ball

   a. is added to the total momentum.
   b. is transferred to the billiard ball.
   c. is transferred to the table holding the balls up.
   d. stays with the cue ball.

19. The law of conservation of momentum states that any time two or more objects interact, they may exchange momentum, but the

total amount of momentum stays the same. True or False?
(Circle one.)

Review (p. 153)

Now that you've finished Section 2, review what you've learned by answering the Review questions in your ScienceLog.

CHAPTER

7 ▏ **DIRECTED READING WORKSHEET**

Forces in Fluids

As you read Chapter 7, which begins on page 160 of your textbook, answer the following questions.

Imagine . . . (p. 160)

1. The _____ is the only place on Earth deep enough to swallow the tallest mountain in the world.

2. How might the design of *Deep Flight* allow it to go to such depths and move quickly through the ocean?

_____

_____

_____

_____

_____

What Do You Think? (p. 161)

Answer these questions in your ScienceLog now. Then later, you'll have a chance to revise your answers based on what you've learned.

Investigate! (p. 161)

3. How do you think this activity would demonstrate the effect of pressure on a fluid?

_____

_____

_____

Section 1: Fluids and Pressure (p. 162)

4. How are dogs, flies, dolphins, and humans connected by fluids?

_____

_____

5. What else can a fluid do besides flow?

_____

_____

CHAPTER 7 ▲ ▲ ▲

6. What can particles in a fluid do that particles in a solid can't do?

   a. They can stay rigidly in place.
   b. They can melt.
   c. They can move easily past each other.
   d. None of the above

All Fluids Exert Pressure (p. 162)

7. Why does a tire expand as you pump air into it?

   _____

   _____

   _____

   _____

   _____

8. To calculate pressure, _____ the force exerted by a fluid by the _____ over which the force is exerted. (multiply or divide, area or volume)

9. Can you blow a bubble that is cube-shaped? Explain.

   _____

   _____

   _____

   _____

   _____

Atmospheric Pressure (p. 163)

Choose the item in Column B that best matches the phrase in Column A, and write your answer in the space provided.

Column A	Column B
____ **10.** pressure caused by the weight of the atmosphere	**a.** 10
____ **11.** percentage of gases found within 10 km of Earth's surface	**b.** atmospheric pressure
____ **12.** holds the atmosphere in place	**c.** 80
____ **13.** number of newtons pressing on every square centimeter of your body	**d.** gravity

---

Chapter 7, continued

---

14. The depth of a fluid is related to the pressure it exerts. The deeper you go in a fluid, the _____ the pressure becomes. (lower or greater)

15. Look at Figure 4 and number the following locations in order from lowest (1) to highest (5) pressure.

_____ Mount Everest's peak

_____ La Paz, Bolivia

_____ Airplane at cruising altitude

_____ Malibu Beach

_____ 150,000 m above sea level

16. Why do your ears "pop" as atmospheric pressure decreases?

_____

_____

_____

_____

_____

Review (p. 164)

Now that you've finished the first part of Section 1, review what you learned by answering the Review questions in your ScienceLog.

Water Pressure (p. 165)

17. _____ pressure and

_____ pressure are the two kinds of pressure that contribute to the total pressure on an underwater object.

18. Would you feel more pressure 5 m underwater in a pool or 2 m underwater in a lake? Explain.

_____

_____

_____

_____

19. Water is denser than air. True or False? (Circle one.)

Chapter 7, continued

Fluids Flow from High Pressure to Low Pressure (p. 166)

20. How do you use pressure to sip a drink through a straw?

_____

_____

_____

_____

_____

_____

Read each of the following statements, and describe what happens under the conditions described.

21. Pressure is lower inside the lungs than outside the lungs.

_____

22. Pressure is higher inside a tube of toothpaste than outside the tube.

_____

23. Pressure is higher inside a soda can than outside the can.

_____

Pascal's Principle (p. 167)

24. How does a pumping station affect your shower?

_____

_____

_____

_____

25. Liquids transmit pressure _____ efficiently

than gases do because liquids compress _____
easily than gases. (more or less, more or less)

26. A hydraulic brake system in a car acts as a force multiplier because the pistons that push the brake pads are

_____ than the piston that is pushed by the brake pedal. (larger or smaller)

Review (p. 167)

Now that you've finished Section 1, review what you learned by answering the Review questions in your ScienceLog.

Section 2: Buoyant Force (p. 168)

1. How does the buoyant force cause the rubber duck to float?

_____

_____

_____

Buoyant Force Is Caused by Differences in Fluid Pressure (p. 168)

2. Buoyant force exists because the pressure in a fluid is greater at

the _____ of an object. (top or bottom)

3. What did Archimedes figure out about the buoyant force of an object?

_____

_____

_____

4. Buoyant force is determined by
   a. the weight of the displaced water.
   b. the volume of the displaced water.
   c. the weight of the object.
   d. the volume of the object.

Weight vs. Buoyant Force (p. 169)

5. If the weight of the water an object displaces is equal to the

weight of the object, the object _____.
(floats or sinks)

6. If the weight of the water an object displaces is less than the

weight of the object, the object _____.
(floats or sinks)

CHAPTER 7

Choose the statement in Column B that best matches the description
in Column A, and write the corresponding letter in the space provided.

Column A	Column B
____ **7.** A rock sinks to the bottom of a pond.	**a.** buoyant force < weight
____ **8.** A duck is buoyed to the surface of a pond.	**b.** buoyant force = weight
____ **9.** A fish floats in a pond.	**c.** buoyant force > weight

An Object Will Float or Sink Based on Its Density (p. 170)

10. How does the density of a rock affect its ability to float?

_____

_____

11. Why don't most substances float in air if they float in water?

_____

_____

12. The helium balloon in Figure 11 floats in air because the displaced air is _____ than the helium.
(lighter or heavier)

The Mystery of Floating Steel (p. 171)

13. How does the shape of a steel ship allow the ship to float?

_____

_____

_____

_____

14. When you flood a submarine's tanks with sea water
  a. the submarine becomes less dense and rises.
  b. the submarine becomes less dense and sinks.
  c. the submarine becomes denser and rises.
  d. the submarine becomes denser and sinks.

15. How does a fish's swim bladder cause the fish to move like a submarine?

_____

_____

_____

_____

_____

Review (p. 172)

Now that you've finished Section 2, review what you learned by answering the Review questions in your ScienceLog.

Section 3: Bernoulli's Principle (p. 173)

1. What happens to your shower curtain when you increase the water pressure in your shower?

_____

_____

Fluid Pressure Decreases as Speed Increases (p. 173)

2. What did Bernoulli say about the speed of a moving fluid?

   a. The faster the speed is, the higher the pressure.
   b. The slower the speed is, the lower the pressure.
   c. The faster the speed is, the lower the pressure.
   d. Speed and pressure are not related.

3. The table-tennis ball in Figure 14 stays in the water stream. Why?

_____

_____

_____

It's a Bird! It's a Plane! It's Bernoulli's Principle! (p. 174)

4. Look at Figure 15. The shape of an airplane wing causes the air

above the wing to flow _____ than the air below it. (slower or faster)

5. The upward force acting on an airplane wing due to air flow is

called _____ . (buoyancy or lift)

Chapter 7, continued

Thrust and Wing Size Determine Lift (p. 175)

6. How does thrust increase lift?

_____

_____

_____

7. How large must the wings be for each of the following airplane types? (Circle one for each type.)

    a. high-performance jet   small   medium   large

    b. glider   small   medium   large

8. Which of the methods below do birds use to stay in the air? (Circle all that apply.)

    a. They use large wing size to glide on wind currents.

    b. They pull up their legs.

    c. They move their tails up and down.

    d. They flap their wings.

Drag Opposes Motion in Fluids (p. 176)

9. In a strong wind, drag is the _____ that you walk against.

10. _____ usually causes drag forces in flight.

11. Wing flaps are parts of a commercial airplane that reduce drag.

    True or False? (Circle one.)

Wings Are Not Always Required (p. 177)

12. Take a moment to examine Figure 19. A baseball with a side spin

    will curve _____ the side where the ball is moving in the same direction as the air flow.
(away from or toward)

Review (p. 177)

Now that you've finished Section 3, review what you learned by answering the Review questions in your ScienceLog.

CHAPTER

8 **DIRECTED READING WORKSHEET**

Work and Machines

As you read Chapter 8, which begins on page 186 of your textbook, answer the following questions.

Would You Believe . . . ? (p. 186)

1. The Egyptians built the Great Pyramid in 30 years. Why is this so amazing?

_____

_____

_____

2. Which of the following are simple machines used by the Egyptians to build pyramids?
   a. screws and screwdrivers
   b. inclined planes and levers
   c. plows and axes
   d. wheels and axles

What Do You Think? (p. 187)

Answer these questions in your ScienceLog now. Then later, you'll have a chance to revise your answers based on what you've learned.

Investigate! (p. 187)

3. What is the purpose of this activity?

_____

_____

_____

Section 1: Work and Power (p. 188)

4. In the scientific sense, you are doing work on this page by reading this question. True or False? (Circle one.)

The Scientific Meaning of *Work* (p. 188)

5. When you bowl, you are applying a _____ to the bowling ball that makes it move down the lane.

▲ **CHAPTER 8**

6. When you are carrying a heavy suitcase at a constant speed, you are not doing work. Why not?

_____

_____

_____

_____

7. Look at the chart on page 189. Which of the following is having work done on it? (Circle all that apply.)

a. a grocery bag as you pick it up
b. a grocery bag as you carry it at a constant speed
c. a crate as you push it at a constant speed
d. a backpack as you are walking with a constant speed

Calculating Work (p. 190)

8. Suppose you want to calculate how much work it takes to lift a 160 N barbell. Besides the weight of the barbell, what other information do you need to know?

a. the shape of the weights
b. how high the barbell is being lifted
c. the strength of the person doing the lifting
d. None of the above

9. In the equation for work, F is the _____

applied to the object and d is the _____
through which the force is applied.

Power—How Fast Work Is Done (p. 191)

10. The _____ is the unit used to express power.

11. If you increase power, you are increasing the amount of work

done in a given amount of time. True or False? (Circle one.)

Review (p. 191)

Now that you've finished Section 1, review what you learned by answering the Review questions in your ScienceLog.

Section 2: What Is a Machine? (p. 192)

1. The tools used to fix a flat tire are not complicated enough to be called machines. True or False? (Circle one.)

Machines—Making Work Easier (p. 192)

2. Which of the following are machines? (Circle all that apply.)

   a. scissors   **c.** screw

   b. lug nut   **d.** car jack

Mark each of the following statements *True* or *False*.

3. _____ As you pry the lid off a can of paint, work is done on the screwdriver and on the lid.

4. _____ While you pry off the lid with the screwdriver, you are not doing any work.

5. _____ The force you apply to the screwdriver while prying is the input force.

6. _____ The work output done by the screwdriver works against the forces of weight and friction to open the lid.

7. _____ The screwdriver helps you open the can because it increases the amount of work you apply to the can.

8. How does using a ramp make lifting a heavy object easier?

   a. The object is moved over a shorter distance.
   b. The ramp increases the amount of work you do.
   c. Less force is needed to move the object over a longer distance.
   d. None of the above

9. When a machine shortens the distance over which a force is

   exerted, the size of the force must _____ .
   (increase or decrease)

Mechanical Advantage (p. 196)

10. By comparing the mechanical advantage of two machines, you can tell which machine

    a. is bigger.
    b. has a larger output force.
    c. has a larger input force.
    d. makes work easier.

CHAPTER 8

Chapter 8, continued

---

11. Chopsticks allow you to exert force over a longer distance, but the mechanical advantage is

 a. less than one. **c.** greater than one.
 b. equal to one. **d.** impossible to determine.

Mechanical Efficiency (p. 197)

12. The work output of a machine is always less than the work input. Where does the missing work go?

 a. It is used to get the machine started.
 b. It is used to overcome the friction created by using the machine.
 c. It is used to keep the machine running.
 d. None of the above

13. A machine that has no friction to overcome is called

 a. an ideal machine. **c.** a real machine.
 b. a complex machine. **d.** a smooth machine.

Review (p. 197)

Now that you've finished Section 2, review what you learned by answering the Review questions in your ScienceLog.

Section 3: Types of Machines (p. 198)

1. Name the six simple machines.

_____

_____

Levers (p. 198)

Mark each of the following statements *True* or *False*.

2. _____ A first-class lever changes the direction of the input force.

3. _____ The output force of a second class lever is smaller than the input force.

4. _____ Third-class levers do not increase the input force.

Inclined Planes (p. 200)

5. You must divide the _____ of the

inclined plane by the _____ you are lifting the load in order to calculate mechanical advantage.

6. An inclined plane saves the work required to lift an object.

True or False? (Circle one.)

_____

Chapter 8, continued
_____

Wedges (p. 201)

7. Which of the following is NOT a wedge?

    a. knife   **c.** chisel

    b. plow   **d.** ramp

8. How do you calculate the mechanical advantage of a wedge?

_____

_____

Screws (p. 201)

9. Which of the following is NOT a screw?

    a. jar lid   **c.** bolt

    b. steering wheel   **d.** All three are screws.

10. Like using an inclined plane, using a screw enables you to apply

a _____ force over

_____ distance.

Use what you've learned in the first part of Section 3 to answer
the following questions. Choose the machine in Column B that best
matches the definition in Column A, and write the corresponding
letter in the space provided.

Column A	Column B
_____ **11.** a straight, slanted surface	**a.** screw
_____ **12.** a bar that pivots at a fixed point	**b.** inclined plane
_____ **13.** a double inclined plane that moves	**c.** lever
_____ **14.** an inclined plane wrapped in a spiral	**d.** wedge

Review (p. 202)

Now that you've finished the first part of Section 3, review what you
learned by answering the Review questions in your ScienceLog.

Wheel and Axle (p. 202)

15. In a wheel and axle, which is larger: the radius of the wheel or
the radius of the axle?

_____

16. The input force of a wheel and axle is exerted along

    a. a circular distance.

    b. a rectangular distance.

    c. an inclined plane.

    d. a spiral.

Pulleys (p. 203)

17. Which of the following is NOT true of pulleys?

    a. A pulley is a grooved wheel that holds a rope or cable.

    b. A movable pulley moves up with the load as it is lifted.

    c. Fixed and movable pulleys working together form a block and tackle.

    d. Fixed pulleys increase force.

Compound Machines (p. 204)

18. A compound machine consists of two or more

_____ .

19. A can opener is a compound machine. Which of the following simple machines are part of a can opener? (Circle all that apply.)

    a. lever   **c.** screw

    b. wheel and axle   **d.** wedge

20. A machine with many moving parts generally has a lower mechanical efficiency than a machine with fewer moving parts.

    True or False? (Circle one.)

Review (p. 205)

Now that you've finished Section 3, review what you learned by answering the Review questions in your ScienceLog.

CHAPTER

9   DIRECTED READING WORKSHEET

Energy and Energy Resources

As you read Chapter 9, which begins on page 212 of your textbook, answer the following questions.

Strange but True! (p. 212)

1. What vast treasure-troves have been buried at sea for millions of years?

a. gold   **c.** salt

b. gas hydrates   **d.** sodium bicarbonate

2. Scientists suspect that large areas off the coasts of North Carolina

and South Carolina may contain _____ times the natural gas consumed by the United States in 1 year.

3. What happens when you hold a flame near icy formations of water and methane?

_____

_____

What Do You Think? (p. 213)

Answer these questions in your ScienceLog now. Then later, you'll have a chance to revise your answers based on what you've learned.

Investigate! (p. 213)

4. What will you find out in this activity?

_____

_____

Section 1: What Is Energy? (p. 214)

5. Where do you think energy is being transferred as the tennis game is played?

___ _____

_____

_____

Energy and Work—Working Together (p. 214)

6. Energy is the _____ to do work.

7. When you hit a tennis ball with a racket, energy is transferred from the racket to the ball. True or False? (Circle one.)

8. Work and energy are both measured in

_____ .

Kinetic Energy Is Energy of Motion (p. 215)

9. Does the tennis player have kinetic energy if she isn't moving? Explain.

_____

_____

_____

_____

10. In Figure 2, on page 215, swinging a hammer gives the hammer energy to do work. True or False? (Circle one.)

11. Which of the following have kinetic energy?
(Circle all that apply.)
  a. a falling raindrop   **c.** a plane in the sky
  b. a rolling bowling ball   **d.** a parked car

12. Which of the following is NOT true of kinetic energy?
  a. The faster something moves, the more kinetic energy it has.
  b. The lower the mass is, the higher the kinetic energy.
  c. Speed has a greater effect on kinetic energy than mass has.

13. The truck and the red car in Figure 3, on page 215, are traveling at the same speed. So why does the truck have more kinetic energy?

_____

_____

Potential Energy Is Energy of Position (p. 216)

14. Why does a stretched bow have potential energy?

_____

_____

_____

Chapter 9, continued

---

15. Take a moment to look at Figure 5, on page 216. Which of the following would have more gravitational potential energy than a diver on a platform? (Circle all that apply.)

    a. a diver with the same mass on a lower platform

    b. a diver with the same mass on a higher platform

    c. a diver with more mass on the same platform

    d. a diver with less mass on the same platform

16. What two measurements do you multiply together to get gravitational potential energy?

_____

_____

Mechanical Energy Sums It All Up (p. 217)

17. The mechanical energy of the juggler's pins in Figure 6 is the

total energy of motion and position of the pins. True or False? (Circle one.)

18. Potential energy plus gravitational energy equals mechanical

energy. True or False? (Circle one.)

Review (p. 217)

Now that you've finished the first part of Section 1, review what you learned by answering the Review questions in your ScienceLog.

Forms of Energy (p. 218)

19. List the six forms of energy.

_____

_____

20. The total potential energy of all the particles in an object is

known as thermal energy. True or False? (Circle one.)

21. In Figure 7, on page 218, the particles in ocean water have less kinetic energy than the particles in steam. Why?

_____

_____

_____

_____

CHAPTER 9

Choose the type of energy in Column B that best matches the definition in Column A, and write the corresponding letter in the space provided. The type of energy may be used more than once.

Column A	Column B
____ **22.** energy produced by vibrations of electrically charged particles	**a.** chemical
____ **23.** energy of a compound that changes when its atoms are rearranged to form a new compound	**b.** electrical
____ **24.** energy caused by an object's vibrations	**c.** light
____ **25.** energy of moving electrons	**d.** sound
____ **26.** energy used in radar systems	

27. Nuclear energy can be produced only by splitting the nucleus of an atom. True or False? (Circle one.)

28. Where does the sun get its energy to light and heat the Earth?

_____

_____

_____

29. The nucleus of an atom can store _____ energy. (potential or kinetic)

Review (p. 221)

Now that you've finished Section 1, review what you learned by answering the Review questions in your ScienceLog.

Section 2: Energy Conversions (p. 222)

1. When you are hammering a nail, what is one type of energy conversion that is taking place?

_____

_____

_____

2. An energy conversion can happen between any two forms of energy. True or False? (Circle one.)

From Kinetic to Potential and Back (p. 222)

Take a look at Figure 14, on page 222. Mark each of the following energy conversions $K{\rightarrow}P$ (kinetic to potential) or $P{\rightarrow}K$ (potential to kinetic).

3. _____ You jump down, and the trampoline stretches.

4. _____ The trampoline does work on you, and you bounce up.

5. _____ You reach the top of your jump on the trampoline.

6. _____ You are about to hit the trampoline again.

7. In Figure 15, on page 223, the potential energy of the pendulum is the smallest at which point in its swing?

    a. the highest point   **c.** the slowest point
    b. the lowest point   **d.** none of the above

Conversions Involving Chemical Energy (p. 223)

8. Why does eating breakfast give you energy to start the day?

_____

_____

_____

_____

9. The energy you get from food originally comes from the sun.

True or False? (Circle one.)

10. During _____ , plants convert light energy into chemical energy.

Conversions Involving Electrical Energy (p. 225)

11. Look at Figure 18. Which of the following forms of energy are converted from electrical energy when you turn on a hair dryer? (Circle all that apply.)

    a. nuclear energy   **c.** kinetic energy
    b. sound energy   **d.** thermal energy

12. In a battery, _____ energy is converted

into _____ energy.

Review (p. 225)

Now that you've finished the first part of Section 2, review what you learned by answering the Review questions in your ScienceLog.

CHAPTER 9

Energy and Machines (p. 226)

13. How does a machine make work easier? (Circle all that apply.)

    a. by changing the direction of the required force
    b. by changing the size of the required force
    c. by requiring no force
    d. by increasing the amount of energy transferred

14. A nutcracker can transfer more energy to a nut than you transfer to the nutcracker. True or False? (Circle one.)

15. Which of the following kinetic energy transfers does NOT occur when you ride a bike as in Figure 20 on page 226?

    a. from legs to pedals   **c.** from chain to back wheel
    b. from pedals to chain   **d.** from gear wheel to chain

16. An energy conversion makes the telephone a useful machine. Explain.

_____

_____

_____

17. As gasoline burns inside an engine, _____ energy is converted into thermal and kinetic energy. (electrical or chemical)

Why Energy Conversions Are Important (p. 228)

18. Take a look at Figure 22. How could wind help you cook a meal?

_____

_____

_____

_____

19. The more efficient the light bulb is, the more electrical energy is converted into light energy instead of thermal energy.

True or False? (Circle one.)

20. Energy output is always more than energy input. True or False? (Circle one.)

Review (p. 228)

Now that you've finished Section 2, review what you learned by answering the Review questions in your ScienceLog.

Section 3: Conservation of Energy (p. 229)

1. A roller-coaster car never returns to its starting height because

energy gets lost along the way. True or False? (Circle one.)

Where Does the Energy Go? (p. 229)

2. Where does friction oppose motion on a roller-coaster car?
(Circle all that apply.)

   a. between the wheels of the car and the track
   b. between the car and the surrounding air
   c. between the car's axles and the wheels
   d. between the car and the passenger

3. The original amount of potential energy of a roller-coaster car is

converted into kinetic energy and _____
energy as the car races down the hill. (electrical or thermal)

4. The potential energy of the car at the top of the second hill of a
roller coaster is equal to the original potential energy of the car

at the top of the first hill. True or False? (Circle one.)

Energy Is Conserved Within a Closed System (p. 230)

5. A roller coaster is involved in a closed system. List its parts.

_____

6. The law of conservation of energy states that energy can be neither

_____ nor _____ .

7. Look at Figure 24 on page 230. Which forms of energy are converted from the electrical energy that enters the light bulb?
(Circle all that apply.)

   a. light energy
   b. thermal energy warming the bulb
   c. thermal energy caused by friction in the wire
   d. chemical energy

No Conversion Without Thermal Energy (p. 231)

8. During an energy conversion, energy is rarely converted to

thermal energy. True or False? (Circle one.)

9. Why is it impossible to make a perpetual motion machine?

_____

_____

_____

_____

_____

_____

_____

Review (p. 231)
Now that you've finished Section 3, review what you learned by answering the Review questions in your ScienceLog.

Section 4: Energy Resources (p. 232)
1. How do we use energy resources?

_____

_____

_____

_____

2. The _____ is the energy resource responsible for most other energy resources.

Nonrenewable Resources (p. 232)
3. Which of the following are fossil fuels? (Circle all that apply.)

    a. coal   **c.** petroleum
    b. wood   **d.** natural gas

4. Fossil fuels are formed from the remains of plants and animals

that lived millions of years ago. True or False? (Circle one.)

5. Explain why fossil fuels are concentrated forms of the sun's energy.

_____

_____

_____

_____

_____

_____

Use the images on page 233 to answer questions 6–8.

6. Most of the coal supply in the United States is used for heating. True or False? (Circle one.)

7. Which of the following is NOT a petroleum product?
   a. rayon clothing   **c.** petrochemicals
   b. a candle   **d.** wool

8. The cleanest-burning fossil fuel is _____.

9. Take a moment to study Figure 27 on page 234. Put the following events in the correct sequence for the production of electrical energy from fossil fuels by writing the appropriate number in the space provided.

_____ A large magnet spins within a ring of wire coils.

_____ Thermal energy converts liquid water to steam.

_____ Electric current is generated in the wire coils.

_____ Fossil fuels are burned.

_____ Steam pushes against the blades of a turbine.

_____ Electrical energy is distributed to homes.

_____ Water is pumped into a boiler.

10. In nuclear fission, a nucleus of a(n) _____ element releases energy when it is split into two nuclei.

11. Nuclear energy is considered to be a renewable resource.

   True or False? (Circle one.)

12. One uranium fuel pellet in Figure 28, on page 235, contains the

   energy equivalent of about one _____ of coal.

CHAPTER 9

_____

Chapter 9, continued

_____

Renewable Resources (p. 235)

13. Solar energy can be used to run a television. Explain.

_____

_____

_____

_____

14. Electrical energy generated from falling water is called hydro-gravity. True or False? (Circle one.)

15. A wind turbine converts the _____ energy of the wind into electrical energy. (potential or kinetic)

16. Geothermal energy results from the heating of Earth's

_____ .

17. Which of the following is NOT an example of biomass?

    a. leaves   **c.** wood

    b. steel   **d.** cactus

18. Corn can be used to make a cleaner-burning fuel for cars.

True or False? (Circle one.)

The Two Sides to Energy Resources (p. 237)

Choose the energy resource in Column B that best matches the disadvantage in Column A, and write the corresponding letter in the space provided.

Column A	Column B
____ **19.** requires large areas of farmland	**a.** fossil fuels
____ **20.** produces radioactive waste	**b.** nuclear
____ **21.** requires dams that disrupt river ecosystems	**c.** solar
____ **22.** expensive for large-scale energy production	**d.** water
____ **23.** waste water can damage soil	**e.** wind
____ **24.** only practical in windy areas	**f.** geothermal
____ **25.** burning produces smog and acid precipitation	**g.** biomass

Review (p. 237)

Now that you've finished Section 4, review what you learned by answering the Review questions in your ScienceLog.

Heat and Heat Technology

As you read Chapter 10, which begins on page 244 of your textbook, answer the following questions.

Strange but True! (p. 244)

1. How do Earthship windows maximize the sun's radiant energy?

_____

_____

2. What materials are recycled and used for insulation in the walls of an Earthship?

_____

_____

What Do You Think? (p. 245)

Answer these questions in your ScienceLog now. Then later, you'll have a chance to revise your answers based on what you've learned.

Investigate! (p. 245)

3. Before you do this activity, which material, if any, do you think will feel the warmest?

_____

_____

_____

Section 1: Temperature (p. 246)

4. Usually when you turn on the hot-water knob, the water that comes out of the faucet isn't hot at first. Can you think of another time when something is labeled "hot" or "cold" and really isn't?

_____

_____

_____

_____

Chapter 10, continued

What Is Temperature? (p. 246)

5. Temperature is a measure of the average _____ of the particles in an object.

   a. potential energy
   b. mechanical energy
   c. kinetic energy
   d. size

6. The faster the particles of an object are moving, the

   _____ the temperature of the object.
   (higher or lower)

7. All of the particles of a substance at a certain temperature move

   at the same speed and in the same direction. True or False?
   (Circle one.)

8. The temperature of a substance depends on how much of the

   substance you have. True or False? (Circle one.)

Measuring Temperature (p. 248)

9. When a substance undergoes thermal expansion
   (Circle all that apply.)

   a. its particles expand.
   b. its particles spread out.
   c. its particles get colder.
   d. its volume increases.

10. All substances will expand equally with the same change in

    temperature. True or False? (Circle one.)

11. According to the Brain Food on page 248, the hottest tempera-
    ture ever recorded was taken in the Libyan desert. It was 58°C in
    the shade! What is this temperature on the Fahrenheit scale?

    a. 192°F
    b. 89°F
    c. 136°F
    d. 104°F

12. At what temperature does water boil? Write your answer using all
    three of the temperature scales on page 248.

    _____

    _____

13. It is incorrect to say the temperature of an object is 23°K. Why?

    _____

    _____

    _____

_____

Chapter 10, continued
_____

14. What temperature scale do weather reporters in the United States use to tell you how hot it is outside?

   a. Kelvin   **c.** Celsius

   b. Fahrenheit   **d.** SI

15. If you are given a temperature in degrees Fahrenheit and asked to convert it to degrees Celsius, which equation would you use?

   a. $°F = \left(\dfrac{9}{5} \times °C\right) + 32$ **c.** $°C = \dfrac{5}{9} \times (°F - 32)$

   b. $K = °C + 273$ **d.** $°C = K - 273$

More About Thermal Expansion (p. 250)

16. Expansion joints are used in bridges in order to prevent the bridge from

   a. swaying.   **c.** buckling.

   b. getting too hot.   **d.** making noise.

17. The thermostat in your home has a _____ strip that coils or uncoils with changes in

_____ .

18. Figure 4, on page 250, shows how a thermostat works. When room temperature falls below the desired temperature, the strip

in your thermostat _____ and an electric

circuit is _____ . (coils or uncoils, opened or closed)

Review (p. 250)

Now that you've finished Section 1, review what you learned by answering the Review questions in your ScienceLog.

Section 2: What Is Heat? (p. 251)

1. A stethoscope feels colder than a tongue depressor when it touches your skin. Why?

_____

_____   _____

Heat Is a Transfer of Energy (p. 251)

2. Under what condition can heat occur between two objects?

   a. The objects must be hot.

   b. The objects must be large.

   c. The objects must be at different temperatures.

   d. The objects must have a lot of energy.

3. If two objects come in contact with each other, and one object is warmer than the other object, what happens?

   a. Both objects get colder.

   b. Both objects get warmer.

   c. Energy is transferred from the colder object to the warmer object.

   d. Energy is transferred from the warmer object to the colder object.

4. If two objects have the same temperature, but the larger object has more moving particles than the smaller object, what do you know about the thermal energy of the two objects?

   a. The larger object has more thermal energy.

   b. The smaller object has more thermal energy.

   c. They have the same thermal energy.

   d. None of the above

5. The temperature of an object _____ as its particles slow down. (decreases or increases)

6. You place a warm bottle of juice in ice water like in Figure 7 on page 252. After the bottle and the ice water have reached thermal equilibrium

   a. the juice is cooler than the water.

   b. the water is cooler than the juice.

   c. the water and the juice have the same temperature.

   d. the water is cooler than it was before.

Conduction, Convection, and Radiation (p. 253)

7. In conduction, when faster-moving particles collide with slower-moving particles, the faster-moving particles

_____ kinetic energy to the slower-moving particles.

8. In the table of conductors and insulators on page 254, what do all of the objects named as conductors have in common?

   a. They are all used for cooking.   **c.** They have the same shape.

   b. They are all made of metal.   **d.** They have the same size.

9. In a convection current, warmer particles

_____ because they are

_____ than cooler particles.

(rise or sink, denser or less dense)

10. You can feel the warmth of a portable heater like the one in Figure 10, on page 255, when you are standing near it because of

_____ . (conduction or radiation)

Chapter 10, continued

---

11. How is the greenhouse effect helpful?

_____

_____

_____

12. How is the greenhouse effect harmful?

_____

_____

_____

Review (p. 255)

Now that you've finished the first part of Section 2, review what you learned by answering the Review questions in your ScienceLog.

Heat and Temperature Change (p. 256)

13. The cloth strap of a seat belt has a _____ specific heat capacity than the metal buckle, so it takes

_____ energy to change the temperature of the metal than it takes to change the temperature of the cloth. (higher or lower, more or less)

14. Look at the Meteorology Connection on page 256. How does the temperature of the ocean affect the temperature of coastal areas?

_____

_____

_____

_____

_____

15. Thermal energy is not related to the number of particles in an object. True or False? (Circle one.)

16. When a substance cools down, the value for heat would be

a _____ number. (positive or negative)

17. You can find the specific heat capacity of a substance by using

a _____ . To calculate the specific heat
capacity of a substance you need to know the amount of energy
transferred by the substance to the water, the test substance's

_____ , and the change in its

_____ .

18. What is the relationship between a calorie and a Calorie?

_____

19. Look at the can in Figure 15, on page 258. When you read the
nutritional information on a package of food, the amount of
energy contained in the food in one serving is given in

   a. Calories.   **c.** teaspoons.
   b. calories.   **d.** joules.

The Differences Between Temperature, Thermal Energy, and Heat (p. 259)

Indicate whether each of the following is a characteristic of heat,
temperature, or thermal energy. In the space provided, write *H* for
heat, *T* for temperature, and *TE* for thermal energy.

20. _____ measure of the average kinetic energy of an object's
particles

21. _____ varies with the mass and temperature of a substance

22. _____ total kinetic energy of a substance's particles

23. _____ does not vary with the mass of a substance

24. _____ transfer of energy between objects at different
temperatures

Review (p. 259)

Now that you've finished Section 2, review what you learned by
answering the Review questions in your ScienceLog.

Section 3: Matter and Heat (p. 260)

1. Why does a frozen juice bar melt before you are done eating it?

_____

_____

_____

_____

_____

Chapter 10, continued

_____

States of Matter (p. 260)

Choose the word in Column B that best matches the description in Column A, and write the corresponding letter in the space provided.

Column A	Column B
_____ **2.** Particles can slide past one another.	**a.** solid
_____ **3.** Particles move independently of one another.	**b.** liquid
_____ **4.** Particles vibrate in place.	**c.** gas

Changes of State (p. 261)

Write the change of state described by each of the following in the appropriate space.

5. liquid changes to gas _____

6. liquid changes to solid _____

7. gas changes to liquid _____

8. solid changes to liquid _____

9. A change of state changes

    a. the chemical properties of a substance.
    b. the melting point of a substance.
    c. the identity of the substance.
    d. None of the above

10. Look at graph on page 261. When you heat liquid water past 100°C, the temperature of the water does not rise above 100°C until all of the water has become steam.

    True or False? (Circle one.)

11. According to the Biology Connection in the upper right of page 262, how do scientists determine the number of Calories in food?

    a. They find the volume of the food sample.
    b. They burn a dry sample in a calorimeter.
    c. They dissolve the food sample in water
    d. They eat a sample of food and see how fast they can run.

Heat and Chemical Changes (p. 262)

12. What occurs in a chemical change?

_____

_____

_____

▲▶ **CHAPTER 10**

13. Energy is always involved when bonds between particles are broken or formed. True or False? (Circle one.)

Review (p. 262)

Now that you've finished Section 3, review what you learned by answering the Review questions in your ScienceLog.

Section 4: Heat Technology (p. 263)

1. Besides a heater, name another example of heat technology.

_____

Heating Systems (p. 263)

2. In a hot-water heating system, what heats the air in the house?

_____

_____

_____

3. In a warm-air heating system, warm air travels from the furnace

through the house in _____ and into

rooms through _____ .

4. If your house has no insulation, what must be done to keep your house warm in the winter?

_____

_____

5. Look at Figure 21 on page 264. Air is a good insulator.

True or False? (Circle one.)

6. Which of the following is NOT a component of a passive solar heating system?
   a. large windows facing south **c.** moving parts
   b. thick walls   **d.** good insulation

7. Where is water heated in an active solar heating system?

_____

_____

_____

_____

Heat Engines (p. 266)

8. In the process of combustion

    a. fuel combines with oxygen to produce thermal energy.

    b. fuel combines with steam to produce thermal energy.

    c. air combines with water to produce steam.

    d. air combines with oxygen to produce steam.

9. A steam engine is an _____ combustion engine. The hot gaseous mixture that expands

to do work is _____ .
(external or internal, steam or fuel)

10. The engine in a car is an _____ combustion engine. (external or internal)

In a car's engine, the pistons make four strokes. Label the following in the order that the strokes are made by writing the appropriate number in the space provided.

11. _____ the compression stroke

12. _____ the power stroke

13. _____ the intake stroke

14. _____ the exhaust stroke

Cooling Systems (p. 267)

15. Cooling systems transfer _____ energy out of a warm area, so that it feels cooler.

16. Thermal energy moves naturally from areas of

_____ temperature to areas of

_____ temperature.

17. A refrigerator is cool inside because it transfers

    a. thermal energy from the condenser coils to the inside of the refrigerator.

    b. specific heat from the condenser coils to the inside of the refrigerator.

    c. thermal energy from inside the refrigerator to the condenser coils.

    d. specific heat from inside the refrigerator to the condenser coils.

Heat Technology and Thermal Pollution (p. 269)

18. Thermal pollution hurts a river's ecosystem by

    a. contaminating the water.

    b. damaging the river's organisms with smog.

    c. heating the river excessively.

    d. introducing new species to the ecosystem.

19. What can power plants do to reduce thermal pollution?

_____

_____

_____

_____

_____

_____

20. Take a moment to read the Environment Connection on page 269. Where do heat islands form? Explain.

_____

_____

_____

_____

_____

Review (p. 269)

Now that you've finished Section 4, review what you learned by answering the Review questions in your ScienceLog.

CHAPTER

11 DIRECTED READING WORKSHEET

Introduction to Atoms

As you read Chapter 11, which begins on page 278 of your textbook, answer the following questions.

Would You Believe . . . ? (p. 278)

1. What do dinosaurs have in common with atoms?

   _____

   _____

2. How did scientists find information that caused them to change their theory about the way *T. rex* walked? (Circle all that apply.)
   a. by studying well-preserved dinosaur tracks
   b. by examining similarities between the skeletons of *T. rex* and an ostrich
   c. by observing a *T. rex* as it was walking
   d. by extracting DNA from fossilized mosquitoes

3. Scientists are able to develop theories about dinosaurs and

   atoms only through _____ evidence.
   (direct or indirect)

What Do You Think? (p. 279)

Answer these questions in your ScienceLog now. Then later, you'll have a chance to revise your answers based on what you've learned.

Investigate! (p. 279)

4. How do you think rolling marbles in this activity will help you identify the mystery object?

   _____

   _____

   _____

   _____

   _____

Section 1: Development of the Atomic Theory (p. 280)

5. Atoms are NOT

    a. a relatively new idea to us.

    b. the building blocks of all matter.

    c. the smallest particles into which an element can be divided and still be the same substance.

    d. seen with the scanning tunneling microscope.

6. An explanation that is supported by testing and brings together a broad range of hypotheses and observations is called a

_____ .

Democritus Proposes the Atom (p. 280)

7. The word *atom* comes from a Greek word that means

_____ . (invisible or indivisible)

8. Which of the following statements is part of Democritus's theory about atoms?

    a. Atoms are small, soft particles.

    b. Atoms are always standing still.

    c. Atoms join together to form different materials.

    d. None of the above

9. We know that Democritus was right to say that all matter was made up of atoms. So why did people ignore Democritus's ideas for such a long time?

_____

_____

_____

Dalton Creates an Atomic Theory Based on Experiments (p. 281)

10. By conducting experiments and making observations, Dalton figured out that elements combine in random proportions because they're made of individual atoms. True or False? (Circle one.)

11. Dalton's theory states that atoms cannot be

_____ , _____ , or

_____ .

12. Atoms of different elements are exactly alike.

True or False? (Circle one.)

13. How did Dalton think atoms formed new substances?

_____

_____

Thomson Finds Electrons in the Atom (p. 282)

Mark the following statements *True* or *False*.

14. _____ In 1897, J. J. Thomson made a discovery that proved the first part of Dalton's atomic theory was correct.

15. _____ Thomson discovered that there were small particles inside the atom.

16. _____ Thomson found that the electrically charged plates affected the direction of a cathode-ray tube beam.

17. _____ Thomson knew the beam was made of particles with a positive charge because it was pulled toward a positive charge.

18. When you rub a balloon on your hair, your hair is

_____ to the balloon because both the

hair and the balloon have become _____ .

19. The two types of charge are positive and neutral.

True or False? (Circle one.)

20. Objects with the same charge attract each other. True or False? (Circle one.)

21. In Thomson's "plum-pudding" model, electrons are NOT
   a. negatively charged.
   b. present in every type of atom.
   c. collected together in the center of the atom.
   d. scattered throughout a blob of positively charged material.

Review (p. 283)

Now that you've finished the first part of Section 1, review what you learned by answering the Review questions in your ScienceLog.

Rutherford Opens an Atomic "Shooting Gallery" (p. 284)

22. Before his experiment, Rutherford expected the particles to

deflect to the sides of the gold foil. True or False? (Circle one.)

23. Review Figure 6 and read the text on page 285. Figure 6 shows the new atomic model resulting from Rutherford's experiment. Which of the following statements is NOT part of Rutherford's revision of his former teacher's atomic theory?

    a. Atoms are made mostly of empty space.
    b. The nucleus is a dense, charged center of the atom.
    c. Lightweight, negative electrons move in the nucleus.
    d. Most of an atom's mass is in the nucleus.

24. The diameter of the nucleus of an atom is

    _____ times smaller than the diameter of

    the atom.

Bohr States That Electrons Can Jump Between Levels (p. 286)

25. In Bohr's atomic model, _____ travel

    in definite paths around the _____ in

    specific levels. Each level is a certain _____
    from the nucleus. Electrons cannot be found between levels, but they can

    _____ from level to level.

26. Bohr's model only predicted some atomic behavior.

    True or False? (Circle one.)

The Modern Theory: Electron Clouds Surround the Nucleus (p. 286)

27. The exact path of a moving electron can now be predicted.

    True or False? (Circle one.)

28. What are electron clouds?

    _____

    _____

    _____

Review (p. 286)

Now that you've finished Section 1, review what you learned by answering the Review questions in your ScienceLog.

Section 2: The Atom (p. 287)

1. In this section you will learn about the particles inside the atom

and the _____ that act on the particles
inside the atom.

How Small Is an Atom? (p. 287)

Each of the following statements is false. Change the underlined
word to make the statement true. Write the new word in the space
provided.

2. A sheet of aluminum foil is about <u>500</u> atoms thick.

_____

3. <u>An Olympic medal</u> contains about twenty thousand billion

billion atoms of copper and zinc. _____

What's Inside an Atom? (p. 288)

Choose the term in Column B that best matches the phrase in
Column A, and write the appropriate letter in the space provided.

Column A	Column B
____ **4.** particle found in the nucleus that has no charge	**a.** electron cloud
____ **5.** particle found in the nucleus that is positively charged	**b.** electron
____ **6.** particle with an unequal number of protons and electrons	**c.** amu
____ **7.** negatively charged particle found outside the nucleus	**d.** nucleus
____ **8.** size of this determines the size of the atom	**e.** proton
____ **9.** contains most of the mass of an atom	**f.** ion
____ **10.** SI unit used for the masses of atomic particles	**g.** neutron

Review (p. 289)

Now that you've finished the first part of Section 2, review what you
learned by answering the Review questions in your ScienceLog.

How Do Atoms of Different Elements Differ? (p. 289)

11. The simplest atom is the _____ atom. It
has one proton and one electron.

Chapter 11, continued

---

12. Neutrons in the atom's nucleus keep two or more protons from

    moving apart. True or False? (Circle one.)

13. If you build an atom using two protons, two neutrons, and
    two electrons, you have built an atom of

    _____ .

14. An element is composed of atoms with the same number of

    _____ . (neutrons or protons)

Are All Atoms of an Element the Same? (p. 290)

15. It is NOT true that isotopes of an element
    a. have the same number of protons but different numbers of
       neutrons.
    b. are stable when radioactive.
    c. share most of the same chemical properties.
    d. share most of the same physical properties.

Calculating the Mass of an Element (p. 292)

16. The weighted average of the masses of all the naturally occurring

    isotopes of an element is called _____ mass.

What Forces Are at Work in Atoms? (p. 293)

Choose the type of force in Column B that best matches the phrase in
Column A, and write the corresponding letter in the space provided.

Column A	Column B
____ **17.** counteracts the electromagnetic force so protons stay together in the nucleus	**a.** gravity
____ **18.** depends on the mass of objects and the distance between them	**b.** electromagnetic force
____ **19.** plays a key role in neutrons changing into protons and electrons in unstable atoms	**c.** strong force
____ **20.** holds the electrons around the nucleus	**d.** weak force

Review (p. 293)

Now that you've finished Section 2, review what you learned by
answering the Review questions in your ScienceLog.

CHAPTER

12 ▍ **DIRECTED READING WORKSHEET**

The Periodic Table

As you read Chapter 12, which begins on page 300 of your textbook, answer the following questions.

Would You Believe . . . ? (p. 300)

1. Hyraxes are related to elephants, even though they don't look alike. What have scientists similarly discovered about different-looking elements?

_____

_____

_____

2. The periodic table is useful for _____ the known elements and predicting the _____ of unknown elements.

What Do You Think? (p. 301)

Answer these questions in your ScienceLog now. Then later, you'll have a chance to revise your answers based on what you've learned.

Investigate! (p. 301)

3. What will you be looking for in this activity?

_____

_____

_____

Section 1: Arranging the Elements (p. 302)

4. Why do you think scientists might have been frustrated by the organization of the elements before 1869?

_____

_____

_____

_____

_____

▲ **CHAPTER 12**

Discovering a Pattern (p. 302)

5. Mendeleev spent a lot of train rides organizing the elements according to their properties. Which arrangement of elements produced a repeating pattern of properties?

    a. by increasing density

    b. by increasing melting point

    c. by increasing shine

    d. by increasing atomic mass

6. How are the days of the week periodic?

_____

_____

_____

_____

7. Mendeleev noticed after arranging the elements that similar

_____ and _____
properties could be observed in every

_____ element.

8. Mendeleev was able to predict the properties of elements that no one knew about. How was this possible?

_____

_____

_____

_____

Changing the Arrangement (p. 303)

9. A few elements in Mendeleev's table seemed to be mysteriously out of place according to their properties. How did Moseley solve the mystery? (Circle all that apply.)

    a. He rearranged the elements by atomic number.

    b. He discovered protons, neutrons, and electrons.

    c. He disproved the periodic law.

    d. He determined the number of protons in an atom.

10. The basis of the periodic table is the periodic

_____ , which states that the properties of

elements are _____ of their atomic

_____ .

_____

Chapter 12, continued

Use the periodic table on pages 304–305 of your text to fill in the answers to the following questions.

11. Which information is NOT included in each square of the periodic table in your text?

a. atomic number   **c.** melting point
b. chemical symbol   **d.** atomic mass

12. How can you tell that chlorine is a gas at room temperature?

_____

_____

_____

_____

_____

13. Rows of elements are called _____ , and

columns of elements are called _____ or

_____ .

14. Who will approve the names of the newest elements?

a. the scientist who discovered each element
b. an international committee of scientists
c. the chemists from a research institute

15. Silicon is a _____ .
(metal, nonmetal, or metalloid)

Finding Your Way Around the Periodic Table (p. 306)

16. The properties of elements determine whether elements are classified as metals, nonmetals, or metalloids. The number of

_____ in the outer

_____ level of an atom helps determine which of these three categories an element belongs to.

17. There is a zigzag line on the periodic table. How can it help you?

_____

_____

_____

Chapter 12, continued

Use the pictures on pages 306–307 to help you match the category in Column B with the description in Column A, and write the corresponding letter in the space provided. Categories may be used more than once.

Column A	Column B
_____ **18.** few electrons in the outer energy level	**a.** metals
_____ **19.** have some properties of the other two categories	**b.** nonmetals
_____ **20.** brittle and nonmalleable solids	**c.** metalloids
_____ **21.** complete or almost-complete set of electrons in the outer energy level	
_____ **22.** conducts heat from a stovetop to your food	
_____ **23.** can prevent a spark from igniting gasoline in your car	
_____ **24.** outer energy level containing a shell of electrons that is about half-complete	
_____ **25.** formed into electrical wires	
_____ **26.** flattened into sheets of food wrap without shattering	
_____ **27.** border the zigzag line on the periodic table	

28. Some elements are named after scientists, like Einstein, and

places, like California. True or False? (Circle one.)

29. The chemical symbol Pb comes from the

_____ word *plumbum*, which means

_____ .

30. What happens as you move from left to right through each period on the periodic table?

a. Elements change from having properties of nonmetals to having properties of metals.

b. Elements change from having properties of metalloids to having properties of metals.

c. Elements change from liquids to gases.

d. None of the above

Review (p. 309)

Now that you've finished Section 1, review what you learned by answering the Review questions in your ScienceLog.

_____

Chapter 12, continued
_____

Section 2: Grouping the Elements (p. 310)

1. Why do elements in a family or group have similar properties?

    a. They have the same atomic mass.

    b. They have the same number of protons in their nuclei.

    c. They have the same number of electrons in their outer energy level.

    d. They have the same number of total electrons.

Groups 1 and 2: Very Reactive Metals (p. 310)

2. The elements in Groups 1 and 2 are very reactive. Explain.

_____

_____

_____

_____

_____

_____

_____

3. Which of the following is NOT true of alkali metals?

    a. They can be cut with a knife.

    b. They are usually stored in water.

    c. They are the most reactive of all the metals.

    d. They can easily give away their outer electron.

4. How are the following alkali metal compounds useful?

    a. sodium chloride _____

    b. sodium hydroxide _____

    c. potassium bromide _____

5. Alkaline-earth metals have _____ electrons in their outer energy level. They are less _____ _____ and more _____ than alkali metals.

6. Calcium is the alkaline-earth metal that makes up a compound that is healthy for your teeth. True or False? (Circle one.)

▲▲ CHAPTER 12

Groups 3–12: Transition Metals (p. 312)

7. Besides collectively being called transition metals, Groups 3–12

also have individual names. True or False? (Circle one.)

8. Which of the following characteristics describe transition metals? (Circle all that apply.)

   a. good conductors
   b. more reactive than alkali and alkaline-earth metals
   c. one or two electrons in the outer energy level
   d. denser than elements in Groups 1 and 2

9. Mercury is different from the other transition metals in Figure 7. How?

_____

_____

10. Two rows of _____ metals are placed at the bottom of the periodic table to save space. Elements in the

_____ row are called lanthanides and are

shiny, _____ metals.

11. Which lanthanide forms a compound that makes you see red on a computer screen like the one in Figure 8?

_____

_____

12. All actinides are radioactive. True or False? (Circle one.)

13. Which actinide is used in some smoke detectors?

_____

_____

Review (p. 313)

Now that you've finished the first part of Section 2, review what you learned by answering the Review questions in your ScienceLog.

_____

Chapter 12, continued

_____

Groups 13–16: Groups with Metalloids (p. 314)

14. Look at Figure 9. The most common element of Group 13, aluminum, was once considered so valuable that Napoleon III used it as dinnerware. True or False? (Circle one.)

15. What do diamonds, crayons, and proteins have in common?

_____

16. Phosphorous, which makes up about 80 percent of the air you breathe, is used in fertilizers. True or False? (Circle one.)

17. All substances need the element oxygen to

_____ .

Complete the following section after you finish reading about Groups 13–16. Each of the following statements is false. Change the underlined word to make the statement true. Write the new word in the space provided.

18. Oxygen group elements contain five electrons in the outer level.

_____

19. The carbon group contains no nonmetals.

_____

20. Nitrogen group and boron group elements vary in reactivity.

_____

21. Not all carbon group and oxygen group elements are solid at room temperature.

_____

Groups 17 and 18: Nonmetals Only (p. 316)

22. Which of the following statements is true?

    a. Group 17 elements are the most reactive metals.
    b. Group 18 elements are the least reactive metals.
    c. Group 18 elements are the least reactive nonmetals.

23. What does Figure 12 show about the physical properties of halogens?

_____

_____

▲ CHAPTER 12

Chapter 12, continued

24. Halogens are very reactive because of the number of electrons in their outer energy level. True or False? (Circle one.)

25. What important use do the halogens iodine and chlorine have in common?

_____

_____

26. Which of the following statements are true of noble gases? (Circle all that apply.)

   a. They are colorless and odorless at room temperature.
   b. They normally react with other elements.
   c. They are metals.
   d. They have a complete set of electrons in their outer energy level.

27. Take a moment to look at Figure 13. Why do neon signs contain other noble gases besides neon? Give an example.

_____

_____

_____

_____

Hydrogen Stands Apart (p. 317)

Mark each of the following statements *True* or *False*.

28. _____ Hydrogen is useful as rocket fuel.

29. _____ Hydrogen has two electrons in its outer energy level.

30. _____ Hydrogen is the most abundant element in the universe.

31. _____ The physical properties of hydrogen are closer to those of nonmetals than to those of metals.

32. _____ Hydrogen does not react with oxygen.

Review (p. 317)

Now that you've finished Section 2, review what you learned by answering the Review questions in your ScienceLog.

CHAPTER

13   DIRECTED READING WORKSHEET

Chemical Bonding

As you read Chapter 13, which begins on page 326 of your textbook, answer the following questions.

Strange but True! (p. 326)

1. A scientist discovered superglue by accident. What was he trying to develop?

_____

_____

_____

2. What type of bond is the force that holds atoms together?

   a. electrical   **c.** gravitational

   b. chemical   **d.** static

3. Which of the following are possible uses of superglue? (Circle all that apply.)

   a. attaching aircraft parts

   b. repairing a cracked tooth

   c. fertilizing plants

What Do You Think? (p. 327)

Answer these questions in your ScienceLog now. Then later, you'll have a chance to revise your answers based on what you've learned.

Investigate! (p. 327)

4. What will you be observing in this activity?

_____

_____

_____

Section 1: Electrons and Chemical Bonding (p. 328)

5. Every substance in the world can be made out of about 100

   elements. True or False? (Circle one.)

▲▲ CHAPTER 13
▲

Chapter 13, continued

Atoms Combine Through Chemical Bonding (p. 328)

6. Sugar is made from atoms of which of the following elements? (Circle all that apply.)

    a. carbon   **c.** hydrogen

    b. nitrogen   **d.** oxygen

7. A chemical bond is the _____ of attraction that holds a pair of atoms together.

Electron Number and Organization (p. 329)

8. In order to make the overall charge of an atom zero, there must be an

equal number of negatively charged _____

and positively charged _____ .

9. Valence electrons are the electrons in an atom's innermost

energy level. True or False? (Circle one.)

Look at Figure 3 on page 330. Write the number of valence electrons for each of the following elements:

10. _____ oxygen

11. _____ sodium

12. _____ chlorine

13. _____ helium

To Bond or Not to Bond (p. 330)

14. Which electrons determine whether or not an atom will form bonds?

    a. the electrons in the nucleus

    b. the electrons in the innermost energy level

    c. the electrons in the outermost energy level

    d. None of the above

15. An atom will not normally form a chemical bond if it has

_____ valence electrons.

16. Which of the following does NOT describe how atoms can fill their outermost energy level?

    a. by sharing electrons with other atoms

    b. by losing electrons to other atoms

    c. by gaining electrons from other atoms

    d. by gaining kinetic energy from other atoms

_____

Chapter 13, continued
_____

17. Why does a helium atom need only two valence electrons?

_____

_____

_____

Review (p. 331)

Now that you've finished Section 1, review what you learned by answering the Review questions in your ScienceLog.

Section 2: Types of Chemical Bonds (p. 332)

1. Three types of chemical bonds are _____ ,

_____ , and _____ .

Ionic Bonds (p. 332)

2. Describe how two atoms can become ions.

_____

_____

_____

3. An atom that loses one or more electrons from its outermost

energy level becomes a positively charged ion. True or False? (Circle one.)

4. Which of the following elements give up electrons to other atoms? (Circle all that apply.)

   a. sodium   **c.** chlorine

   b. aluminum   **d.** oxygen

5. Why do the elements in Groups 1 and 2 react so easily?

_____

_____

_____

6. Atoms of nonmetals lose one or more protons when they form

ionic bonds. True or False? (Circle one.)

7. The names of negative ions that form when atoms gain electrons

have the ending _____ .

8. A large amount of energy is released when atoms of Group 17

elements lose electrons. True or False? (Circle one.)

9. Which of the following are common properties of an ionic
compound? (Circle all that apply.)

   a. Its solid form is a crystal lattice.
   b. It contains alternating positive and negative ions.
   c. It is soft and pliable at room temperature.
   d. Its positive and negative ions repel each other.
   e. It has a low melting point.
   f. It has a high boiling point.
   g. It is neutral.

10. Look at Figure 8 on page 335. What force causes both the forma-
tion of ionic bonds and static cling?

   a. the Earth's gravity
   b. the repulsion of like charges
   c. the attraction of opposite charges
   d. a magnetic pole

Review (p. 335)
Now that you've finished the first part of Section 2, review what you
learned by answering the Review questions in your ScienceLog.

Covalent Bonds (p. 336)
11. Covalent bonds form between atoms that require a large amount

of energy in order to lose an electron. True or False? (Circle one.)

12. In a covalent bond, neither atom loses or gains an electron.

Instead, one or more electrons are _____
by the atoms. (shared or created)

13. Look at Figure 11 on page 336. The electrons that are shared by
two atoms spend most of their time

   a. near the smallest of the two atoms.
   b. near the largest of the two atoms.
   c. between the nuclei of the two atoms.
   d. in the nuclei of the two atoms.

14. A group of atoms held together by covalent bonds is a neutral

particle called a _____ .

15. Draw the electron-dot diagram for water.

16. Draw the electron-dot diagram for krypton.

17. In an electron-dot diagram, each dot represents one proton.

True or False? (Circle one.)

18. Diatomic molecules are the simplest kinds of molecules. They consist of two atoms bonded together. True or False? (Circle one.)

19. Give three examples of complex molecules.

_____

_____

20. Carbon is known as the building block of life. Which of the following is a property of this important element? (Circle all that apply.)

a. Each of its atoms needs to make four bonds.
b. It is found in all proteins.
c. It can bond with other elements and form long chains.
d. It is in a water molecule.

Metallic Bonds (p. 339)

21. In a metal, "swimming" protons surround the metal ions.

True or False? (Circle one.)

22. What are three properties of metals that are a result of metallic bonding?

_____

_____

_____

_____

_____

23. Because ions in a metal can be easily rearranged without break-ing the metallic bonds, metals tend to be easily

    a. shattered.   **c.** reshaped.

    b. crystallized.   **d.** broken.

24. Which of the following is NOT a typical property of a metal?

    a. malleability   **c.** conductivity

    b. ductility   **d.** brittleness

25. Besides being valuable in the jewelry industry, gold is special because it can be hammered into a very thin foil. This property is called

    a. malleability.   **c.** conductivity.

    b. ductility.   **d.** brittleness.

Identify each of the following substances as containing mostly ionic, mostly covalent, or mostly metallic bonds. Refer back to the earlier parts of Section 2 as needed. Write *I* for ionic, *C* for covalent, and *M* for metallic.

26. _____ copper wire

27. _____ water

28. _____ table salt

29. _____ sugar

30. _____ carbon dioxide

31. _____ plaster of Paris

32. _____ aluminum foil

33. _____ gold jewelry

34. _____ sea shells

Review (p. 341)

Now that you've finished Section 2, review what you learned by answering the Review questions in your ScienceLog.

CHAPTER

14 DIRECTED READING WORKSHEET

Chemical Reactions

As you read Chapter 14, which begins on page 348 of your textbook, answer the following questions.

Imagine . . . (p. 348)

The moment an air bag-equipped vehicle slams into a wall, a sequence of events rapidly takes place to protect you from hitting the dashboard. Place the events in the proper sequence by writing the appropriate number in the space provided.

1. ____   A small electric current is sent to the gas generator.

2. ____   The air bag fills the space between you and the dashboard.

3. ____   A sensor detects the sudden decrease in speed.

4. ____   The air bag inflates with gas formed in the gas generator.

5. ____   Chemicals in the gas generator react, creating a gas.

6. Could you construct an air bag using vinegar and baking soda for the gas generator? Explain.

_____

_____

_____

_____

_____

What Do You Think? (p. 349)

Answer these questions in your ScienceLog now. Then later, you'll have a chance to revise your answers based on what you've learned.

Investigate! (p. 349)

7. What is the purpose of this activity?

_____

_____

_____

Section 1: Forming New Substances (p. 350)

8. The color of leaves that contain chlorophyll is

_____ .

▲ ▲ CHAPTER 14
▲

_____

Chapter 14, continued

_____

9. The red, orange, and yellow colors of leaves are always present

but are hidden until the chlorophyll breaks down. True or False? (Circle one.)

Chemical Reactions (p. 350)

10. Chemical reactions do not change the substances involved.

True or False? (Circle one.)

11. Look at Figure 2 on page 350. What causes bubbles to form in a muffin?

_____

_____

_____

_____

Choose the clue to a chemical reaction in Column B that best matches the example in Column A, and write the corresponding letter in the space provided.

Column A	Column B
____ **12.** thermal energy produced by a fire	**a.** color change
____ **13.** precipitate	**b.** energy change
____ **14.** bubbles	**c.** solid formation
____ **15.** white spots caused by bleach	**d.** gas formation

16. What does a chemical reaction have to do with making and breaking chemical bonds?

_____

_____

_____

_____

_____

Chemical Formulas (p. 352)

17. The subscript in the chemical formula H_2O tells you there are two

   a. atoms of hydrogen in the molecule.
   b. electrons on the hydrogen atom in the molecule.
   c. elements in the molecule.
   d. atoms of oxygen in the molecule.

In the space provided, write the number of atoms of each element in each of the following chemical formulas.

18. O_2 _____

19. $C_6H_{12}O_6$ _____

20. H_2O _____

21. $Ca(NO_3)_2$ _____

22. Covalent compounds are usually composed of two or more

   _____ .

23. In the chart on page 353, what number does the prefix *hepta-* stand for?

   a. six   **c.** seven
   b. eight   **d.** five

Write the formula for each of the following covalent compounds.

24. dinitrogen monoxide _____

25. carbon dioxide _____

26. Ionic compounds are composed of a _____

   and a _____ .

27. The overall charge of an ionic compound is zero. True or False? (Circle one.)

Write the formula for each of the following ionic compounds.

28. sodium chloride _____

29. magnesium chloride _____

Chemical Equations (p. 354)

30. What do musical notation and chemical equations have in common?

   _____

   _____

   _____

▲ CHAPTER 14

31. When carbon reacts with oxygen to form carbon dioxide,

carbon dioxide is the _____ .
(product or reactant)

32. Look at Figure 9. In a chemical equation, the formulas of the

_____ appear before the arrow, and

the formulas of the _____ appear after
the arrow.

33. Which of the following are diatomic elements?
(Circle all that apply.)

a. oxygen   **d.** bromine
b. carbon   **e.** nitrogen
c. neon   **f.** hydrogen

34. If you make a mistake in a chemical formula, you may be

describing a very different substance. True or False? (Circle one.)

35. According to the Brain Food on page 355, what is one good
reason to use hydrogen gas as a fuel?

_____

_____

_____

36. The coefficient 2 in $2CO_2$ means that there are

a. two oxygen atoms and one carbon atom present.
b. two oxygen atoms present.
c. two carbon atoms and two oxygen atoms present.
d. two carbon dioxide molecules present.

37. After you finish looking at Figure 11 on page 356, balance the
following chemical equation: $O_2 + H_2 \rightarrow H_2O$.

_____

38. Chemical equations must be balanced. Why?

_____

_____

_____

_____

_____

_____

39. Antoine Lavoisier's work led to the law of

_____ of mass, which states that

mass is neither _____ nor

_____ in chemical or physical changes.

Review (p. 357)

Now that you've finished Section 1, review what you learned by answering the Review questions in your ScienceLog.

Section 2: Types of Chemical Reactions (p. 358)

1. Which of the following is NOT one of the four classifications of chemical reactions discussed in the text?

    a. synthesis   **c.** single-replacement

    b. decomposition   **d.** double-decomposition

Synthesis Reactions (p. 358)

2. In a synthesis reaction, a single compound is formed from two or more substances. True or False? (Circle one.)

Decomposition Reactions (p. 359)

3. Decomposition reactions are the _____ of synthesis reactions.

Single-Replacement Reactions (p. 359)

4. How is a person who cuts in on a dancing couple like a single-replacement reaction?

_____

_____

_____

_____

5. In a single-replacement reaction, a more reactive element can replace a less reactive element from a compound.

True or False? (Circle one.)

_____

_____

Double-Replacement Reactions (p. 360)

6. _____ in two _____
   switch places in a double-replacement reaction.

After reading Section 2, match these reaction types with their correct examples. Choose the type of reaction from Column B that best matches the example in Column A, and write the corresponding letter in the space provided.

Column A	Column B
____ **7.** Zinc reacts with hydrochloric acid to form zinc chloride and hydrogen.	**a.** decomposition
____ **8.** Water can be broken down to form hydrogen and oxygen.	**b.** double-replacement
____ **9.** Magnesium reacts with oxygen in the air to form magnesium oxide.	**c.** single-replacement
____ **10.** Sodium fluoride and silver chloride are formed from the reaction of sodium chloride with silver fluoride.	**d.** synthesis

Review (p. 360)

Now that you've finished Section 2, review what you learned by answering the Review questions in your ScienceLog.

Section 3: Energy and Rates of Chemical Reactions (p. 361)

1. The rate at which a chemical reaction occurs cannot be changed.

   True or False? (Circle one.)

Every Reaction Involves Energy (p. 361)

2. Chemical bonds break as they _____
   energy. (absorb or release)

3. An exothermic reaction releases energy. True or False? (Circle one.)

4. Look at Figure 20 on page 361. Give one example of how energy is released during a chemical reaction.

   _____

   _____

   _____

5. In an endothermic reaction, the chemical energy of the

_____ is less than the chemical energy of

the _____ .

6. What is the law of conservation of energy?

_____

_____

_____

7. What happens to the energy absorbed by endothermic reactions?

_____

_____

8. Which of the following statements about chemical reactions is
NOT true?

   a. Exothermic reactions release energy.
   b. Energy can be stored in molecules.
   c. The activation energy of a chemical reaction is the amount of
   energy released.
   d. All chemical reactions require some energy to get started.

9. Once an exothermic reaction is started, it continues to supply
the activation energy needed for the substances to react.

   True or False? (Circle one.)

Factors Affecting Rates of Reactions (p. 363)

10. What are four factors that affect how rapidly a chemical
reaction takes place?

_____

_____

_____

_____

11. The light stick in Figure 23, on page 364, glows brighter in hot
water because the rate of a reaction

_____ as temperature increases.
(decreases or increases)

12. The concentration of a solution is a measure of the amount of

one substance dissolved in another. True or False? (Circle one.)

13. How does increasing concentration increase the rate of reaction? (Circle all that apply.)

  a. There are more reactant particles present, so the particles are more likely to collide with each other.

  b. The distance between particles is smaller, so the particles are able to collide more frequently.

  c. More particles present gives a lowered activation energy.

  d. Increasing the concentration of a solution lowers the surface area of the reactants.

14. What is one way you can increase the surface area of a solid reactant?

_____

_____

_____

15. A catalyst lowers the activation energy of a reaction.

True or False? (Circle one.)

16. The catalysts used in your body are called

_____ .

17. Which of the following slows down a reaction?

  a. platinum in a catalytic converter

  b. a food preservative

  c. an enzyme in your body

  d. None of the above

18. Which of the following can increase the rate of a chemical reaction? (Circle all that apply.)

  a. raising the temperature

  b. reducing the amount of one substance dissolved in another

  c. increasing the surface area of a solid reactant

  d. adding a catalyst

  e. adding an inhibitor

Review (p. 365)

Now that you've finished Section 3, review what you learned by answering the Review questions in your ScienceLog.

CHAPTER

15 DIRECTED READING WORKSHEET

Chemical Compounds

As you read Chapter 15, which begins on page 372 of your textbook, answer the following questions:

Strange but True . . . (p. 372)

1. What was the inventor of Silly Putty® trying to do?

_____

_____

_____

What Do You Think? (p. 373)

Answer these questions in your ScienceLog now. Then later, you'll have a chance to revise your answers based on what you've learned.

Investigate! (p. 373)

2. The purpose of this activity is to see how the type of

_____ in compounds can determine

the _____ of the compounds.

Section 1: Ionic and Covalent Compounds (p. 374)

3. Which of the following is NOT true about chemical compounds?

    a. Chemical compounds are composed of molecules or ions.
    b. There are actually only a few kinds of chemical compounds.
    c. Chemical compounds have different kinds of bonds.
    d. Chemical compounds are all around us.

Ionic Compounds (p. 374)

4. An ionic bond forms when electrons are transferred from metal

atoms to nonmetal atoms. True or False? (Circle one.)

5. When sodium reacts with chlorine, _____

forms.

6. A crystal lattice is a _____ three-
dimensional pattern of ions.

7. In a crystal lattice, each ion is surrounded by and bonded to ions

with the same charge. True or False? (Circle one.)

8. Circle the properties of ionic compounds.

strong bonds   weak bonds

hard to break   shatter easily

high melting point   low melting point

solid at room temperature   liquid at room temperature

difficult to dissolve in water   easy to dissolve in water

9. _____ ionic compounds cannot conduct
an electric current. (Dissolved or Undissolved)

Covalent Compounds (p. 375)

10. Which of the following statements is NOT true of covalent
compounds?

   a. They form when atoms of two elements share electrons.
   b. They have weaker bonds than ions in a crystal lattice.
   c. They are composed of independent particles called molecules.
   d. They have a higher melting point than ionic compounds.

11. Why don't water and oil mix?

   _____

   _____

   _____

   _____

12. Most solutions that contain molecules of covalent compounds

   do not conduct an electric current. True or False? (Circle one.)

Review (p. 376)

Now that you've finished Section 1, review what you learned by
answering the Review questions in your ScienceLog.

Section 2: Acids, Bases, and Salts (p. 377)

1. Lemon changes the color of tea because the lemon contains a

   substance called a(n) _____. (acid or base)

Chapter 15, continued

Acids (p. 377)

2. Why should you never use taste as a test to identify an unknown chemical?

_____

_____

3. Acids react with all metals to produce hydrogen gas.

True or False? (Circle one.)

4. When an acid is placed in water, the number of hydrogen ions,

H^+, _____ . These extra hydrogen ions

combine with _____ molecules to form

_____ ions.

5. A substance that changes color in the presence of an acid or a

base is called an indicator. True or False? (Circle one.)

6. Blue litmus paper tests for the presence of bases. True or False?
(Circle one.)

Choose the acid in Column B that best matches each use in Column A,
and write the corresponding letter in the space provided. Acids may
be used more than once.

Column A	Column B
____ **7.** paper and paint production	**a.** citric acid
____ **8.** the "bite" in soft drinks	**b.** nitric acid
____ **9.** digestion	**c.** sulfuric acid
____ **10.** in orange juice	**d.** hydrochloric acid
____ **11.** algae preventative	**e.** carbonic acid
____ **12.** rubber production	

13. An acid is strong if all of its molecules break apart in water to

produce hydrogen ions. True or False? (Circle one.)

14. Which of the following are weak acids? (Circle all that apply.)

sulfuric acid   carbonic acid

phosphoric acid   citric acid

nitric acid   hydrochloric acid

Bases (p. 379)

15. Bases have a _____ taste and feel

_____ .

16. Bases increase the number of hydrogen ions in a solution.

True or False? (Circle one.)

Match each of the bases in Column B with the common uses in
Column A, and write the corresponding letter in the space provided.

Column A	Column B
____ **17.** treating heartburn	**a.** ammonia
____ **18.** unclogging drains and making soap	**b.** calcium hydroxide
____ **19.** making cement	**c.** sodium hydroxide
____ **20.** household cleaning	**d.** magnesium hydroxide

21. A base is strong when all the molecules break apart in water to

produce hydrogen ions. True or False? (Circle one.)

Acids and Bases Neutralize One Another (p. 380)

22. How do antacids get rid of heartburn?

_____

23. H^+ ions of an acid and OH^- ions of a base combine to form the

compound we know as _____ .

24. A solution with a pH of 3 is _____ .

(basic or acidic)

25. Which of the following can be used to determine the pH of a
solution? (Circle all that apply.)

    a. a mixture of indicators   **c.** litmus paper
    b. a pH meter   **d.** lemon juice

26. Living things need a steady _____ in
their environment.

27. Why is the helicopter in Figure 15, on page 381, dumping a base
in the lake?

_____

_____

_____

Salts (p. 382)

28. The _____ ion of a base and the

_____ ion of an acid can combine to

form a(n) _____ compound called a salt.

29. Salts can be produced in many different reactions. List the three
reactions shown in Figure 16 on page 382.

_____

_____

_____

Mark each of the following statements *True* or *False*.

30. _____ Calcium chloride is used to season your food.

31. _____ Salt can keep the roads ice-free in winter.

32. _____ The walls of your room may contain calcium sulfate.

33. _____ Sodium nitrate is used in food preservation.

Review (p. 382)

Now that you've finished Section 2, review what you learned by
answering the Review questions in your ScienceLog.

Section 3: Organic Compounds (p. 383)

1. Organic compounds are composed of _____
molecules.

Each Carbon Atom Forms Four Bonds (p. 383)

2. Carbon atoms are able to bond with up to four other atoms.

True or False? (Circle one.)

3. Take a look at Figure 18 on page 383. Which of the following is
NOT a type of carbon backbone?

a. carbon atoms connected in a line
b. a chain continuing in more than one direction
c. a chain forming a ring
d. None of the above

Biochemicals: The Compounds of Life (p. 384)

4. Biochemicals are organic compounds made by living things.

True or False? (Circle one.)

5. What are carbohydrates? (Circle all that apply.)

   a. biochemicals

   b. sources of energy

   c. one or more simple sugars bonded together

Identify each item below as describing simple (S) or complex (C) carbohydrates.

6. _____ many sugar molecules bonded together

7. _____ produced by plants through photosynthesis

8. _____ bread, cereal, and pasta

9. _____ stored, extra sugar

10. _____ one or a few sugar molecules bonded together

11. _____ fruits and honey

12. Which of the following is NOT an example of a lipid?

   a. candle   **c.** sugar

   b. chicken fat   **d.** corn oil

13. Lipids store energy, make up cell membranes, and do not dissolve in water. True or False? (Circle one.)

14. Lipids store excess _____ in the body.

Plants tend to store lipids as _____ .

Animals tend to store lipids as _____ .

Mark each of the following statements about phospholipid molecules *True* or *False*.

15. _____ The molecules help control the movement of chemicals into or out of the cell.

16. _____ The head of a molecule is a long carbon backbone composed of only carbon and hydrogen atoms.

17. _____ Water is attracted to the heads of the molecules.

18. _____ The cell membrane is mostly made of three layers of the molecules.

19. Take a look at the Brain Food in the right column of page 385. Do you think we should eliminate cholesterol from our diets? Explain.

_____

_____

_____

20. Biochemicals composed of _____ are called proteins.

Chapter 15, continued

21. Which of the following are functions of proteins?
(Circle all that apply.)

    a. regulate chemical activities   **c.** transport and store materials
    b. store excess sugar   **d.** provide support

22. Proteins are all the same size and shape. True or False?
(Circle one.)

23. The shape adopted by the bonded amino acids determines the

function of the protein. True or False? (Circle one.)

24. Insulin is one of the smallest proteins. What very important job
does this small protein have?

_____

_____

25. The protein in your blood that carries oxygen to all parts of your

body is called _____ .

26. Some large proteins help control the transport of materials into

and out of cells. True or False? (Circle one.)

27. Why are nucleic acids sometimes called the "blueprints of life?"

_____

_____

_____

28. Nucleic acids are composed of atoms of carbon, phosphorous,

hydrogen, _____ , and

_____ .

Mark each of the following statements *True* or *False*.

29. _____ The two types of nucleic acids are DNA and RNA.

30. _____ RNA is the only genetic material in the cell.

31. _____ DNA molecules can only store a tiny amount of
information due to their small size.

32. _____ RNA contains the information to make DNA.

33. _____ RNA is involved in protein-building.

Review (p. 387)

Now that you've finished the first part of Section 3, review what you
learned by answering the Review questions in your ScienceLog.

Chapter 15, continued

Hydrocarbons (p. 388)

For each of the following statements, identify the hydrocarbons as saturated (S), unsaturated (U), or aromatic (A).

34. _____ At least two carbon atoms share a double or a triple bond.

35. _____ Each carbon atom shares a single bond with four other atoms.

36. _____ Air fresheners and mothballs contain this type of hydrocarbon.

37. _____ No atoms can be added without replacing an atom that is part of the hydrocarbon.

38. _____ Most are based on benzene.

39. _____ Many medicines are manufactured using this type of hydrocarbon.

40. _____ Ethene helps ripen fruit.

41. _____ Propane is this type of hydrocarbon.

Other Organic Compounds (p. 389)

42. All organic compounds are made of only carbon and hydrogen.

True or False? (Circle one.)

Match each organic compound in Column B with one of its uses in Column A, and write the corresponding letter in the space provided.

Column A	Column B
_____ **43.** food preservative	**a.** ester
_____ **44.** fragrance	**b.** alkyl halide
_____ **45.** antifreeze	**c.** organic acid
_____ **46.** refrigerant	**d.** alcohol

Review (p. 389)

Now that you've finished Section 3, review what you learned by answering the Review questions in your ScienceLog.

CHAPTER

16 DIRECTED READING WORKSHEET

Atomic Energy

As you read Chapter 16, which begins on page 396 of your textbook, answer the following questions.

Would You Believe . . . ? (p. 396)

RTGs are useful in places where replacing a battery or using solar power would be difficult. In items 1–3, explain why using solar power is difficult in each of the following places:

1. near Saturn

_____

2. in the deep ocean

_____

3. on the Arctic icecap

_____

What Do You Think? (p. 397)

Answer these questions in your ScienceLog now. Then later, you'll have a chance to revise your answers based on what you've learned.

Investigate! (p. 397)

4. What is the purpose of this activity?

_____

_____

_____

Section 1: Radioactivity (p. 398)

5. Almost everyone has learned something by obtaining unexpected results. Discuss a time when this happened to you or one of your friends.

_____

_____

_____

_____

Chapter 16, continued

Discovering Radioactivity (p. 398)

6. Items that glow when exposed to light are made from

_____ materials.

7. Why did Becquerel place paper over the photographic plate?
   a. so he could protect the plate from scratches
   b. so he could protect the fluorescent mineral from scratches
   c. so light would not reach the plate
   d. so the X rays would not reach the plate

8. _____ was the element in the fluorescent mineral that gave off the invisible energy.

9. Marie Curie used the term *radioactivity* to describe what property?

_____

_____

_____

_____

_____

Nuclear Radiation Is Produced Through Decay (p. 399)

For each type of radiation described in Column A, choose the correct description of what is released in Column B. Write your answer in the space provided.

Column A	Column B
____ **10.** alpha decay	**a.** high-energy light
____ **11.** beta decay	**b.** positron or electron
____ **12.** gamma decay	**c.** helium nucleus

13. Carbon-14 undergoes _____ .

14. Radium-226 undergoes _____ .

The Penetrating Power of Radiation (p. 401)

15. _____ are the most penetrating of the three forms of nuclear radiation.

16. _____ have the greatest mass of the three forms of nuclear radiation.

17. All of the following are symptoms of radiation sickness *except*

   a. hair loss.
   b. burns.
   c. loss of appetite.
   d. rashes.

18. According to the Brain Food on page 401, Marie Curie died of

_____ , which was probably caused by

exposure to _____ .

19. Of the three forms of radiation, _____
do the most damage to living cells.

20. Take a moment to read the Environment Connection on page 402. What happens when you breathe radon-222?

_____

_____

_____

_____

_____

Review (p. 402)

Now that you've finished the first part of Section 1, review what you learned by answering the Review questions in your ScienceLog.

Nuclei Decay to Become Stable (p. 402)

21. The _____ keeps protons and neutrons
together in the nucleus of an atom.

22. All nuclei composed of 83 protons or more are radioactive because the strong force

   a. is weakened by electrons.
   b. gets weaker over time.
   c. acts only at short distances.
   d. cannot overcome repulsion of the neutrons.

23. What does uranium-238 become after a series of 14 decays takes place?

_____

_____

_____

Chapter 16, continued

_____

Finding a Date by Decay (p. 403)

24. From birth, the levels of carbon-14 found in an organism increase greatly as the organism grows older and dies.

True or False? (Circle one.)

25. The rate of decay of a radioactive isotope is called

_____ .

26. Why do you think it is difficult to use carbon-14 to determine the age of extremely old objects (over 50,000 years)?

_____

_____

_____

_____

Radioactivity and Your World (p. 405)

27. How are tracers used? (Circle all that apply.)

    a. detecting tumors without using surgery
    b. removing tumors without using surgery
    c. determining fertilizer quality
    d. dating dinosaur bones

28. In Figure 11, how will the engineer detect a leak?

_____

_____

_____

Review (p. 405)

Now that you've finished Section 1, review what you learned by answering the Review questions in your ScienceLog.

Section 2: Energy from the Nucleus (p. 406)

1. Why do you think it is important for ordinary citizens to know about the benefits and hazards of using nuclear energy?

_____

_____

_____

Nuclear Fission (p. 406)

2. In nuclear fission, an atom decays into two

_____ stable nuclei. (more or less)

3. Compared to the amount of matter in the nucleus, the small amount of energy that is released when a single uranium nucleus

splits is relatively tiny. True or False? (Circle one.)

4. What is a nuclear chain reaction?

_____

_____

_____

_____

5. Nuclear power plants are _____ expensive to operate than conventional power plants. (more or less)

6. The nuclear power plant in Figure 15 uses

_____ nuclei as fuel.

7. Name two hazards associated with the use of nuclear fission to produce power.

_____

_____

_____

Nuclear Fusion (p. 410)

8. Where does fusion occur?
   a. in the Earth's core
   b. on Saturn's surface
   c. in the sun
   d. in modern power plants

9. Look at the Astronomy Connection on page 410. What other substances do stars use as fuel besides hydrogen?

_____

_____

_____

10. What is one reason that today's nuclear power plants do not use fusion reactions to produce power?

_____

_____

_____

_____

_____

11. What kind of fuel have experimental fusion reactors used?

    a. radon-222

    b. hydrogen-3

    c. uranium-235

    d. helium-4

12. Give three reasons why fusion reactors, in theory, would be better than fission reactors.

_____

_____

_____

_____

_____

_____

_____

Review (p. 411)

Now that you've finished Section 2, review what you learned by answering the Review questions in your ScienceLog.

CHAPTER

17 DIRECTED READING WORKSHEET

Introduction to Electricity

As you read Chapter 17, which begins on page 420 of your textbook, answer the following questions.

Strange but True! (p. 420)

1. Which of the following is NOT true about electric eels?

a. An electric eel uses electric discharges to stun or kill its prey.

b. The body of an adult eel can generate 5,000 to 6,000 V.

c. The eel's thick skin protects it from electrocuting itself.

d. Eels swallow their prey whole.

2. The bursts of voltage that an eel gives off when it shocks its prey

is greater than the voltage of an electrical outlet. True or False? (Circle one.)

What Do You Think? (p. 421)

Answer these questions in your ScienceLog now. Then later, you'll have a chance to revise your answers based on what you've learned.

Investigate (p. 421)

3. What is the purpose of this activity?

_____

_____

_____

_____

Section 1: Electric Charge and Static Electricity (p. 422)

4. When you shuffle your feet on the carpet on a dry day, you get a shock from the metal objects that you touch. What is the cause of this?

_____

_____

_____

_____

_____

Atoms and Charge (p. 422)

Choose the word in Column B that best matches the description in
Column A, and write the corresponding letter in the space provided.

Column A	Column B
_____ **5.** composed of atoms	**a.** proton
_____ **6.** a positively charged particle of the nucleus	**b.** neutron
_____ **7.** a negatively charged particle found outside the nucleus	**c.** matter
_____ **8.** a particle of the nucleus that has no charge	**d.** electron

9. According to the law of electric charges, like charges attract and

opposite charges repel. True or False? (Circle one.)

10. Why don't electrons fly out of atoms while traveling around the
nucleus?

_____

_____

11. Which of the following does NOT determine the strength of an
electric force between charged objects?
    a. the age of the charges
    b. the size of the charges
    c. the distance between the charges
    d. All of the above contribute.

12. The region around a charged particle that can exert a force on

another charged particle is called the _____.

Charge It! (p. 424)

Choose the word in Column B that best matches the definition in
Column A, and write the corresponding letter in the space provided.

Column A	Column B
_____ **13.** "wiping" electrons off of one object onto another	**a.** induction
_____ **14.** transfer of electrons when one object touches another object	**b.** friction
_____ **15.** rearranging electrons in an uncharged object when it is near a charged object	**c.** conduction

_____

Chapter 17, continued

_____

16. When objects are charged, charges cannot be created or

destroyed. True or False? (Circle one.)

17. An electroscope can determine which of the following?

    a. whether or not an object is charged
    b. the material that a charged object is made of
    c. the strength of the charge on an object
    d. how many electrons are involved in the charge

Review (p. 426)

Now that you've finished the first part of Section 1, review what you
learned by answering the Review questions in your ScienceLog.

Moving Charges (p. 426)

18. Electric cords are often covered in plastic and have metal prongs.

This is because metal is a good _____ and

plastic is a good _____ .

19. Hair dryers should not be used near water. Why?

_____

_____

_____

_____

Static Electricity (p. 427)

20. What is static electricity?

    a. an electric charge on a stationary object
    b. random electric signals from your dryer
    c. the buildup of electric charges on an object
    d. electricity that moves away from an object

21. As charges move off an object, the object loses its static electri-

city. This process is called electric _____

22. How does clothing that has become charged in a dryer lose its
charge?

_____

_____

23. An electric discharge can occur quickly or slowly. True or False?
(Circle one.)

▲ **CHAPTER 17**

24. Standing on a beach or golf course during a thunderstorm can make you like a lightning rod. Why?

_____

_____

_____

_____

_____

Review (p. 429)

Now that you've finished Section 1, review what you learned by answering the Review questions in your ScienceLog.

Section 2: Electrical Energy (p. 430)

1. Name something that uses electrical energy that would be difficult for you to live without. Explain.

_____

_____

_____

Batteries Are Included (p. 430)

Choose the word in Column B that best matches the definition in Column A, and write the corresponding letter in the space provided.

Column A	Column B
_____ **2.** converts chemical energy into electrical energy	**a.** electrode
_____ **3.** type of cell that contains a solid or pastelike electrolyte	**b.** cell
_____ **4.** mixture of chemicals in a cell	**c.** electrolyte
_____ **5.** where charges enter or exit a cell	**d.** wet
_____ **6.** type of cell that contains a liquid electrolyte	**e.** dry

Bring on the Potential (p. 431)

7. In a battery, electric current exists between the two electrodes

because there is a difference in _____
between the electrodes.

8. If the potential difference is increased, the current

_____ . (increases or decreases)

Other Ways of Producing Electrical Energy (p. 432)

Choose the term in Column B that best matches the description in
Column A, and write the corresponding letter in the space provided.

Column A		Column B
_____ **9.** converts kinetic energy into electrical energy		**a.** silicon
_____ **10.** converts light energy into electrical energy		**b.** copper
_____ **11.** ejects electrons when struck by light		**c.** thermocouple
_____ **12.** converts thermal energy into electrical energy		**d.** photocell
_____ **13.** one type of wire used in some thermocouples		**e.** generator

CHAPTER 17

Review (p. 432)

Now that you've finished Section 2, review what you learned by
answering the Review questions in your ScienceLog.

Section 3: Electric Current (p. 433)

1. Where does most of the electrical energy used in your home come
from?

   a. large rechargeable batteries
   b. electric power plants
   c. chemical reactions
   d. small generators

Current Revisited (p. 433)

2. In a wire, electrons travel at almost the speed of light.

   True or False? (Circle one.)

Label each of the following as a characteristic of alternating current
or direct current. Write *AC* for alternating current or *DC* for direct
current.

3. _____ It's produced by batteries.

4. _____ It provides the energy in your home.

5. _____ The flow of charges switches directions.

6. _____ Charges flow in one direction.

7. _____ It's more practical for transferring electrical energy.

Voltage (p. 434)

8. Another word for potential difference is _____

which is expressed in _____ .

9. As voltage increases in a circuit, the current decreases.

True or False? (Circle one.)

10. The electrical outlets in the United States usually supply a

voltage of _____ V.

11. Look at the Biology Connection on page 435. Why do
doctors intentionally create a large voltage across the chest of
a heart-attack victim?

_____

_____

Resistance (p. 435)

12. Which of the following associations is false?

    a. Increasing resistance increases current.
    b. Decreasing resistance increases current.
    c. Good conductors have low resistance.
    d. Poor conductors have high resistance.

13. An object's resistance depends on which of the following
properties of the object? (Circle all that apply.)

    a. thickness
    b. length
    c. temperature
    d. color

14. The light bulb in Figure 18 has a filament made of tungsten.
Why is tungsten used in this bulb?

_____

_____

_____

15. Thin wires have _____ resistance than

thick wires. Short wires have _____ resistance than long wires.

16. The resistance of metals generally increases with increasing temperatures because at higher temperatures, the faster-moving
atoms slow down the flow of electric charge.

True or False? (Circle one.)

17. In the equation for Ohm's law, what do the letters *I, V,* and *R* stand for?

_____

18. Who was Georg Ohm?

   a. an electrician   **c.** an inventor

   b. a teacher   **d.** an author

19. If you know the current produced in a wire and the voltage applied, you can calculate the resistance of the wire.

   True or False? (Circle one.)

Electric Power (p. 437)

20. Electric power is expressed in

   a. ohms.   **c.** amperes.

   b. volts.   **d.** watts.

21. Light bulbs may be labeled "100 W" or "40 W." This describes

   a. how long they burn.

   b. how they are disposed of.

   c. how fast the light travels.

   d. how bright they glow.

22. A television uses more power than a hair dryer.

   True or False? (Circle one.)

Measuring Electrical Energy (p. 438)

23. In the equation for electrical energy, what do *E, P,* and *t* stand for?

_____

24. What do electric meters measure?

   a. power   **c.** current

   b. voltage   **d.** energy

Review (p. 439)

Now that you've finished Section 3, review what you learned by answering the Review questions in your ScienceLog.

Section 4: Electric Circuits (p. 440)

1. An electric circuit always begins and ends in the same place.

   True or False? (Circle one.)

Parts of a Circuit (p. 440)

2. Which of the following are parts of all electric circuits?
(Circle all that apply.)

    a. a load   **c.** wires

    b. an energy source   **d.** a switch

3. A _____ opens and closes a circuit.

Types of Circuits (p. 441)

4. A _____ circuit has all parts connected in

a single loop. A _____ circuit has different loads on separate branches.

5. A string of holiday lights wired together in series has a burned-
out bulb. Why do all of the lights go out?

_____

_____

_____

6. If a break occurs in one of the loops of a parallel circuit, current

will not run through the other loops. True or False? (Circle one.)

Household Circuits (p. 444)

7. Which of the following may cause a circuit failure?
(Circle all that apply.)

    a. water   **c.** too many loads

    b. broken wires   **d.** excess insulation

8. As more loads are added to a parallel circuit, the entire circuit

draws more current. True or False? (Circle one.)

9. How does a fuse disrupt the flow of charges when the current is
too high?

    a. A metal strip warms up and bends away from the circuit wires.

    b. A metal strip warms up and melts, leaving a gap.

    c. A metal strip changes from a conductor to an insulator.

    d. None of the above

10. Circuit breakers are inconvenient because breakers must be

replaced when they are tripped. True or False? (Circle one.)

Review (p. 445)

Now that you've finished Section 4, review what you learned by
answering the Review questions in your ScienceLog.

CHAPTER

18 **DIRECTED READING WORKSHEET**

Electromagnetism

As you read Chapter 18, which begins on page 452 of your textbook, answer the following questions.

Would You Believe . . . ? (p. 452)

1. How do scientists use magnets to make a frog float in midair?

_____

_____

_____

_____

_____

What Do You Think? (p. 453)

Answer these questions in your ScienceLog now. Then later, you'll have a chance to revise your answers based on what you've learned.

Investigate! (p. 453)

2. When you drag a bar magnet down a nail, what must you be sure to do?

_____

_____

Section 1: Magnets and Magnetism (p. 454)

3. Which of the following statements is NOT true about magnets?

    a. Magnets can stick to some objects without touching them.

    b. Magnets can stick to all types of metals.

    c. Magnets can stick to other magnets.

    d. Magnets are sometimes strong enough to hold up paper on a refrigerator door.

Properties of Magnets (p. 454)

4. _____ , a mineral that attracts objects containing iron, was found by the Greeks in a part of Turkey called Magnesia over _____ years ago.

5. The poles of a magnet are the parts of a magnet where the magnetic effects are strongest. True or False? (Circle one.)

6. If you move the south poles of two magnets together, what will happen? Explain.

_____

_____

7. Take a look at Figure 4 on page 456. What do iron filings do when you sprinkle them around a magnet?

_____

_____

What Makes Materials Magnetic? (p. 456)

8. In some materials, the magnetic fields of the atoms are so

_____ that they group together in small, magnetlike regions called domains. (weak or strong)

9. According to the Brain Food at the bottom left of page 456, how do ranchers protect their cows with magnets?
   a. The cows wear magnets around their necks.
   b. The magnets are ground into a powder and applied to the cows' hides.
   c. The magnets are attached to a barbed wire fence.
   d. The cows swallow magnets that remain in their stomachs.

10. Why would a magnet lose its magnetic properties if it were dropped?

_____

_____

11. When you hold a magnet close to an unmagnetized object, the

_____ in the object align to create a

_____ magnet.
(atoms or domains, temporary or permanent)

12. When you cut a magnet in half, you end up with one north-pole piece and one south-pole piece. True or False? (Circle one.)

Review (p. 458)

Now that you've finished the first part of Section 1, review what you learned by answering the Review questions in your ScienceLog.

Types of Magnets (p. 458)

Choose the type of magnet in Column B that best matches the description in Column A, and write your answer in the space provided.

Column A	Column B
_____ **13.** easy to create but loses its magnetism easily	**a.** ferromagnet
_____ **14.** difficult to magnetize but retains its magnetic properties well	**b.** electromagnet
_____ **15.** produced by an electric current	**c.** temporary magnet
_____ **16.** made from iron, nickel, or cobalt	**d.** permanent magnet

Earth as a Magnet (p. 459)

Mark each of the following statements *True* or *False*.

17. _____ Earth's magnetic poles are not the same as Earth's geographic poles.

18. _____ The magnetic field lines around Earth are very different from those around a bar magnet.

19. _____ The magnetic south pole is at the North Pole.

20. _____ The auroras occur where Earth's magnetic field bends inward at the magnetic poles.

Review (p. 461)

Now that you've finished Section 1, review what you learned by answering the Review questions in your ScienceLog.

Section 2: Magnetism from Electricity (p. 462)

1. Why does a maglev train float?

_____

_____

The Discovery of Electromagnetism (p. 462)

2. Oersted discovered that a wire carrying a(n)

_____ can move a compass needle.

3. The direction of a magnetic field depends on the direction of the

_____ that produced it.

CHAPTER 18

Using Electromagnetism (p. 463)

4. How can you strengthen a solenoid's magnetic field?

_____

_____

_____

5. What makes up an electromagnet?

_____

_____

6. Some electromagnets are strong enough to lift a car.

True or False? (Circle one.)

Magnetic Force and Electric Current (p. 465)

7. A magnet causes a current-carrying wire to move by exerting a

_____ on the wire.

Applications of Electromagnetism (p. 465)

8. How does a solenoid cause your doorbell to ring?

_____

_____

_____

_____

9. An electric motor changes electrical energy into

_____ energy. (kinetic or static)

Mark each of the following statements *True* or *False.*

10. _____ A commutator is a loop or coil of wire in an electric motor that can rotate.

11. _____ Devices used by electricians may contain a galvanometer, which measures voltage.

Review (p. 467)

Now that you've finished Section 2, review what you learned by answering the Review questions in your ScienceLog.

Chapter 18, continued

Section 3: Electricity from Magnetism (p. 468)

1. What process do power companies use to supply your home with electrical energy?

_____

_____

_____

Electric Current from a Magnetic Field (p. 468)

2. In his setup using an iron ring, Faraday discovered that a very strong electromagnet could induce an electric current in the

second wire. True or False? (Circle one.)

3. When did the pointer on Faraday's galvanometer move?

    a. the instant the galvanometer was connected to or disconnected from the battery

    b. as long as the galvanometer was connected to the battery

    c. as long as the galvanometer was disconnected from the battery

4. What process did Faraday discover during his experiment with the iron ring? Explain.

_____

_____

_____

_____

_____

5. Look at Figure 21 on page 469. Which of the following actions would induce a larger current when a magnet moves through a coil of wire?

    a. moving the magnet more slowly

    b. adding more loops of wire

    c. using thicker wire

    d. reversing the direction the magnet moves

6. An electric current is produced in a wire when the wire moves

_____ magnetic field lines.

(across or along)

Chapter 18, continued

Applications of Electromagnetic Induction (p. 470)

7. A generator converts kinetic energy into

_____ energy. (electrical or mechanical)

8. Generators in power plants produce direct current. True or False? (Circle one.)

9. Using the text on page 471, place the following steps of creating electrical energy from nuclear energy in order of occurrence. Write the appropriate number in the space provided.

_____ An electric current is induced, producing electrical energy.

_____ Steam turns a turbine.

_____ A nuclear reaction occurs and creates thermal energy.

_____ The generator's magnet is turned.

_____ Water is boiled to produce steam.

10. What's the difference between a step-up transformer and a step-down transformer?

_____

_____

_____

_____

_____

_____

11. Electric current is distributed to your house at the same voltage that it is produced by power plants. True or False? (Circle one.)

Review (p. 473)

Now that you've finished Section 3, review what you learned by answering the Review questions in your ScienceLog.

CHAPTER

19 **DIRECTED READING WORKSHEET**

Electronic Technology

As you read Chapter 19, which begins on page 480 of your textbook, answer the following questions.

Would You Believe . . . ? (p. 480)

1. What are three differences between the first televisions and modern televisions?

_____

_____

_____

_____

2. What are two advances you'd like to see in the TVs of the future?

_____

_____

_____

What Do You Think? (p. 481)

Answer these questions in your ScienceLog now. Then later, you'll have a chance to revise your answers based on what you've learned.

Investigate! (p. 481)

3. The purpose of this activity is to make a model of a

_____ .

4. In this activity, do you think the length of the string matters? Explain.

_____

_____

_____

Section 1: Electronic Components (p. 482)

5. Electronic devices change electrical energy into thermal energy.

True or False? (Circle one.)

Chapter 19, continued

Inside an Electronic Device (p. 482)

6. Figure 1 shows the _____ inside a TV remote control.

    a. motor   **c.** circuit board

    b. buttons   **d.** TV antenna

7. An LED, or _____ , is one of the electronic components within a TV remote control.

8. The LED in a remote control

    a. gives off radio waves.

    b. is only for decoration.

    c. sends information to the TV.

    d. receives information from the TV.

9. Electronic components control electric current. True or False? (Circle one.)

Semiconductors (p. 483)

10. If a substance conducts _____ energy

better than a(n) _____ but not as well as a conductor, then the substance is called a semiconductor.

11. In pure silicon,

    a. no valence electrons are shared between atoms.

    b. there are no free electrons to create electric current.

    c. there are some atoms of gallium.

    d. there are ten valence electrons.

12. Use Figure 3 to explain why arsenic-doped silicon conducts electric current better than pure silicon.

_____

_____

_____

_____

_____

_____

_____

13. A(n) _____ semiconductor is a semiconductor with "holes." (n-type or p-type)

_____

Chapter 19, continued

_____

Diodes and Transistors (p. 484–486)

After reading pp. 484–486, decide whether each of the following statements describes a diode or a transistor. In the space provided, write *D* if the statement describes a diode and *T* if it describes a transistor.

14. _____ has three "legs" that conduct electric current

15. _____ can be used as a switch

16. _____ a two-layer semiconductor sandwich

17. _____ can be used to amplify electric current

18. _____ can be used to convert AC to DC

19. _____ allows current in only one direction

20. _____ a three-layer semiconductor sandwich

21. Many transistors and diodes can be contained in one integrated

circuit. True or False? (Circle one.)

22. How have integrated circuits affected the size of electronic devices? Explain.

_____

_____

_____

_____

_____

_____

_____

23. Which of the following statements is NOT true about vacuum tubes?

   a. They were invented before transistors.
   b. They could not amplify electric current.
   c. They can be used to change AC to DC.
   d. They were used in early radios.

▲ CHAPTER 19

24. What are two reasons transistors have replaced vacuum tubes in many modern electronic devices?

_____

_____

_____

_____

_____

_____

Review (p. 487)

Now that you've finished Section 1, review what you learned by answering the Review questions in your ScienceLog.

Section 2: Communication Technology (p. 488)

1. The first telecommunication device was the

a. telephone.   **c.** fax.

b. television.   **d.** telegraph.

2. Using the table in Figure 10, on page 488, write your name in International Morse Code in the space provided below.

_____

_____

_____

Communicating with Signals (p. 488)

3. A signal is anything, such as a sound or a series of numbers, that

represents _____ .

4. A signal may travel better inside a _____ ,
which is another form of energy.

Analog Signals (p. 489)

5. Read the Geology Connection. Since a seismogram is an analog signal, it shares none of the properties of the waves produced by

an earthquake. True or False? (Circle one.)

6. Your telephone's analog signal is a wave of

_____ . (sound or electric current)

7. Use Figure 11 to place the following events in the correct order to explain how a telephone works. Write the appropriate number in the space provided.

_____ Electric current is converted to a sound wave.

_____ Vibrations are converted to electric current.

_____ Sound waves enter the transmitter and vibrate a metal disk.

_____ An analog signal travels through the phone wires.

8. In vinyl records, the number and depth of the contours in the grooved walls represent the _____ and

_____ of the sounds.

9. The stylus riding in the grooves of a record causes an electromagnet to vibrate. True or False? (Circle one.)

Digital Signals (p. 490)

Choose the word or phrase in Column B that best matches the description in Column A, and write your answer in the space provided.

Column A	Column B
_____ **10.** represented by a pulse	**a.** a digital signal
_____ **11.** represented by a missing pulse	**b.** binary
_____ **12.** a series of electrical pulses representing the digits of binary numbers	**c.** the number 0
_____ **13.** means "two"	**d.** the number 1

14. Why are the pits and lands on a CD important?

_____

_____

_____

15. The higher the sampling rate, the better the digital signal reproduces the original sound wave. True or False? (Circle one.)

16. Why do CDs last so long?

_____

_____

_____

Radio and Television (p. 492)

17. Waves that travel as changing electric and magnetic fields are

called _____ waves.

18. Look at Figure 16. A modulator
   a. changes sound waves into electric current.
   b. strengthens the analog signal.
   c. combines the analog signal with radio waves.
   d. removes the radio waves from the analog signal.

Mark each of the following statements *True* or *False*.

19. _____ Electrons are accelerated toward your TV screen by special guns.

20. _____ In television, only audio signals are carried by radio waves.

21. _____ Fluorescent materials glow red, blue, and green when beams of electrons strike the TV screen.

Review (p. 493)

Now that you've finished Section 2, review what you learned by answering the Review questions in your ScienceLog.

Section 3: Computers (p. 494)

1. The alarm clock pictured in Figure 18 is a computer. True or False? (Circle one.)

What Is a Computer? (p. 494)

2. Computers can think. True or False? (Circle one.)

3. Which of the following are the basic functions performed by a computer? (Circle all that apply.)
   a. processing   **d.** output
   b. input   **e.** programming
   c. deleting   **f.** storage

Historic Developments (p. 495)

4. Which of the following statements is NOT true of the ENIAC?

   a. It was built with thousands of vacuum tubes.
   b. It was very inexpensive to build.
   c. It was the first general-purpose computer.
   d. It was developed by the United States Army.

5. Integrated circuits have made today's computers small and

   powerful. True or False? (Circle one.)

Computer Hardware (p. 496)

Read pp. 496–497. Then choose the part of the computer in
Column B that best matches the example or definition in Column A,
and write your answer in the space provided. Some parts may be
used more than once.

Column A	Column B
____ **6.** a monitor	**a.** input device
____ **7.** where information is given to a computer	**b.** central processing unit
____ **8.** a keyboard	
____ **9.** where the computer does calculations	**c.** memory
____ **10.** RAM	**d.** output device
____ **11.** where a computer stores data	
____ **12.** where a computer shows its results	
____ **13.** your voice	

14. What is the difference between RAM and ROM?

   _____

   _____

   _____

   _____

   _____

15. What do computers use to "talk" with each other?

   a. modems   **c.** CD-ROMs
   b. speakers   **d.** scanners

Chapter 19, continued

Computer Software (p. 498)

16. Software is a set of _____ that makes it possible for a computer to perform a task.

Mark each of the following statements *True* or *False*.

17. _____ Operating system software interprets commands from the input device.

18. _____ A word processor is an example of a utility.

19. _____ Application software supervises the interactions of the software and the hardware.

The Internet—A Global Network (p. 499)

20. Internet connections at home and at school are exactly the same.

    True or False? (Circle one.)

21. Imagine that you are at home communicating over the Internet with someone in Rome, Italy. Describe how you are connected.

    _____

    _____

    _____

    _____

    _____

    _____

Review (p. 499)

Now that you've finished Section 3, review what you learned by answering the Review questions in your ScienceLog.

Name _____ Date _____ Class_____

The Energy of Waves

As you read Chapter 20, which begins on page 508 of your textbook, answer the following questions.

This Really Happened! (p. 508)

1. How did a land-based earthquake cause the worst marine disaster in the history of the town of Kodiak?

_____

_____

_____

_____

_____

_____

_____

What Do You Think? (p. 509)

Answer these questions in your ScienceLog now. Then later, you'll have a chance to revise your answers based on what you've learned.

Investigate! (p. 509)

2. What is the purpose of this activity?

_____

_____

_____

Section 1: The Nature of Waves (p. 510)

3. Which of the following waves might your family have experienced after a day at the beach? (Circle all that apply.)
   a. water waves
   b. microwaves
   c. light waves
   d. sound waves

Waves Carry Energy (p. 510)

4. A wave can carry energy away from its source. True or False? (Circle one.)

Chapter 20, continued

5. Take a moment to examine Figure 1. Floating birds and boats bob up and down and travel in the same direction as waves.

   True or False? (Circle one.)

6. Air doesn't travel with sound waves. Give an example of what would happen if it did.

   _____

   _____

   _____

7. Waves can transfer _____ through the

   vibration of _____ in solid, liquid, or
   gaseous media.

8. If you put an alarm clock inside a jar and remove all the air from the jar, you wouldn't hear the alarm ringing. Why?

   _____

   _____

   _____

   _____

   _____

9. Electromagnetic waves are waves that require a medium to

   transfer energy. True or False? (Circle one.)

10. Take a moment to look at the Astronomy Connection in the left column of page 512. In what way is looking at the light from a star like viewing the past?

   _____

   _____

   _____

   _____

11. Which of the following lists of waves contains ONLY mechanical waves?

   a. radio waves, X rays, sound waves
   b. seismic waves, water waves, microwaves
   c. microwaves, water waves, X rays
   d. sound waves, seismic waves, water waves

Types of Waves (p. 513)

After reading pages 513–515 in your text, match each type of wave in Column B to the correct statements in Column A, and write the corresponding letter in the appropriate space. Wave types can be used more than once.

Column A	Column B
____ **12.** Particles move forward at the crest and backward at the trough.	**a.** longitudinal wave
____ **13.** Particles vibrate up and down.	**b.** transverse wave
____ **14.** Electromagnetic waves are this type of wave.	**c.** surface wave
____ **15.** A rarefaction is a section of this type of wave that is less crowded than normal.	
____ **16.** Particles vibrate back and forth along the path the wave travels.	
____ **17.** A sound wave is this type of wave.	
____ **18.** Two types of waves can combine to form this type of wave.	
____ **19.** Particles travel perpendicular to the direction the wave travels.	
____ **20.** This wave occurs at or near the boundary of two media.	

Review (p. 515)

Now that you've finished Section 1, review what you learned by answering the Review questions in your ScienceLog.

Section 2: Properties of Waves (p. 516)

1. Waves made by the breeze were very different than waves created by the speedboat. Describe the difference.

_____

_____

_____

_____

Chapter 20, continued

Amplitude (p. 516)

2. Which of the following is NOT true of amplitude?

  a. The smaller the amplitude, the more energy that is carried by a wave.

  b. It is the maximum distance a wave vibrates from its resting position.

  c. The larger the amplitude of a water wave, the taller the wave.

  d. All of the above are true.

Wavelength (p. 517)

3. Wavelength can be measured between corresponding points on

two adjacent waves. True or False? (Circle one.)

Frequency (p. 518)

4. The frequency of a wave is the _____ of waves produced in a given amount of time. To measure frequency of a transverse wave, I can count the number of

_____ or _____

that pass a point in a certain amount of time. I can also count

the number of _____ or

_____ for a longitudinal wave.

Mark each of the following statements *True* or *False*.

5. _____ The frequency of a wave is not related to its wavelength.

6. _____ When wavelength decreases, frequency increases.

7. _____ A wave with a low frequency carries less energy than a wave with a high frequency.

8. _____ Frequency is expressed in hertz (Hz).

Wave Speed (p. 519)

9. The speed of a wave is the _____ traveled

by a wave in a given amount of _____ .

Wave speed is equal to _____ times

_____ .

---

10. Complete the MathBreak in your ScienceLog. Then answer the following questions:

a. What can you calculate if you know the frequency and speed of a wave?

_____

_____

b. If you know the speed of a wave, and you want to find its wavelength, what other piece of information would be helpful for you to know?

_____

_____

_____

Review (p. 519)

Now that you've finished Section 2, review what you learned by answering the Review questions in your ScienceLog.

Section 3: Wave Interactions (p. 520)

1. The moon doesn't produce light like the stars do. So why does the moon shine?

_____

_____

Reflection (p. 520)

2. Reflection is when waves hit a barrier and some of them pass through it. True or False? (Circle one.)

3. Water waves cannot be reflected. True or False? (Circle one.)

4. Sound waves that reflect off canyon walls or classroom walls are called _____.

Refraction (p. 521)

5. Why does the pencil in a half-filled glass of water look like it's broken?

_____

_____

_____

_____

6. A wave bends when it enters a new medium because the part of the wave that enters first is traveling at a different speed than the rest of the wave. True or False? (Circle one.)

Diffraction (p. 521)

7. Which of the following are true of diffraction? (Circle all that apply.)

  a. Waves curve or bend when they reach the edge of an object.
  b. Waves bend through an opening.
  c. Sound travels around buildings.
  d. The size of the barrier does not affect the amount of diffraction.

8. Diffraction of a wave depends on what two things?

_____

_____

_____

9. Look at Figure 18 on page 522. If the opening is larger than the wavelength, _____ diffraction occurs.
(a little or a lot of)

Interference (p. 522)

10. Which of the following is TRUE about overlapping waves?

  a. They share the same space.
  b. They cannot be in the same place at the same time.
  c. They pass around each other.

11. Interference is the result of two or more waves combining to form three waves. True or False? (Circle one.)

_____

Chapter 20, continued

Match each type of interference in Column B to the correct statements in Column A, and write the corresponding letter in the appropriate space. Interference types can be used more than once.

Column A	Column B
____ **12.** The crests of two waves overlap. ____ **13.** The resulting wave has a smaller amplitude than the original waves had. ____ **14.** Waves of the same amplitude cancel each other out. ____ **15.** The result is a wave with deeper troughs and higher crests than the original waves. ____ **16.** Crests of one wave overlap with the troughs of another wave.	**a.** constructive interference **b.** destructive interference

17. A standing wave occurs from interference between the original wave and the reflected wave. True or False? (Circle one.)

18. In a standing wave, _____ interference causes portions of the wave to have a large amplitude, while total _____ interference causes other portions of the wave to be at rest.

19. Standing waves only *look* like they're standing still. What's really going on?
   a. Waves are vibrating faster than light.
   b. Waves are no longer vibrating.
   c. Waves are traveling in both directions.

20. Place the following steps in the correct sequence to explain how the marimba player in Figure 23 uses resonance. Write the appropriate number in the space provided.

   _____ The amplitude of the vibrations is increased.

   _____ The bar vibrates.

   _____ The marimba player strikes a bar.

   _____ A loud note is produced.

   _____ The air in the column underneath the bar absorbs energy from the vibrating bar and begins to vibrate.

   _____ The frequency of the air column matches the frequency of the bar.

   _____ The air column resonates with the bar.

21. During resonance, a vibrating object will cause a second object to vibrate when it reaches the second object's resonant frequency.

True or False? (Circle one.)

22. How did the Tacoma Narrows Bridge earn the nickname Galloping Gertie?

   a. The bridge was used most often by people traveling on horseback.

   b. The bridge experienced wavelike motions during strong winds.

   c. The bridge was built by Gertrude Stein in July 1940.

   d. Soldiers marched across it in such a way that it collapsed in 1831.

23. How did resonance contribute to the destruction of Galloping Gertie?

_____

_____

_____

_____

_____

_____

_____

_____

_____

_____

_____

Review (p. 525)

Now that you've finished Section 3, review what you learned by answering the Review questions in your ScienceLog.

CHAPTER

21 **DIRECTED READING WORKSHEET**

The Nature of Sound

As you read Chapter 21, which begins on page 532 of your textbook, answer the following questions.

Would You Believe . . . ? (p. 532)

1. What did Marco Polo see and hear in Asia that amazed him so much?

_____

_____

_____

2. Which of the following are true about the noises made by the sands? (Circle all that apply.)

a. They can be heard more than 50 km away.

b. People have compared the sound to foghorns, moaning, and cannon fire.

c. They don't occur in the United States.

d. They tend to be found in large deserts.

What Do You Think? (p. 533)

Answer these questions in your ScienceLog now. Then later, you'll have a chance to revise your answers based on what you've learned.

Investigate! (p. 533)

3. What is the purpose of this activity?

_____

_____

_____

Section 1: What Is Sound? (p. 534)

4. What are two sounds you hear indoors?

_____

_____

Sound Is Produced by Vibrations (p. 534)

5. The complete back-and-forth motion of an object is called a

_____ .

Look at Figure 1 to answer items 6 and 7.

6. In a compression, the molecules in the air are

_____ closely packed than in the surrounding air. (more or less)

7. In a rarefaction, the molecules in the air are

_____ closely packed than in the surrounding air. (more or less)

8. Sound waves are an example of _____ waves. (transverse or longitudinal)

9. Take a moment to read the Biology Connection on page 535. What causes the sounds that you make when you speak? (Circle all that apply.)

   a. air making your vocal cords vibrate
   b. air rushing down your windpipe
   c. air vibrating in your chest
   d. air forced through your windpipe

10. Air does not travel with sound waves. But what would happen at the school dance if air did travel with sound?

_____

_____

Creating Sounds Vs. Detecting Sound (p. 535)

11. What happens to the surrounding air when a tree falls and hits the ground?

_____

_____

_____

Sound Waves Require a Medium (p. 536)

12. In the example above, the medium through which sound travels is the ground. True or False? (Circle one.)

13. If a tree fell in a vacuum, why wouldn't you hear a sound?

_____

_____

_____

14. Take a moment to read the Astronomy Connection on page 536. Name one kind of wave that can travel in a vacuum.

_____

How You Detect Sound (p. 536)

15. After your ears convert sound waves into electrical signals, where are the signals sent for interpretation?

a. pinna   **c.** brain

b. spinal cord   **d.** oval window

16. Place each of the following terms in the correct column: *hammer, stirrup, oval window, cochlea, hair cells, pinna, eardrum, ear canal, anvil.* (If the term refers to an entrance to a part of the ear, write the term between the appropriate columns.)

Outer ear		Middle ear		Inner ear

Choose the term in Column B that best matches the description in Column A, and write your answer in the space provided.

Column A	Column B
_____ **17.** the outermost portion of the ear	**a.** cochlea
_____ **18.** bends to stimulate nerves	**b.** pinna
_____ **19.** portion of the ear that contains liquid	**c.** hammer
_____ **20.** the bone that vibrates the oval window	**d.** stirrup
_____ **21.** the eardrum makes this bone vibrate	**e.** hair cell

22. What ways can you protect yourself from tinnitus? (Circle all that apply.)

a. drinking a glass of milk each day

b. wearing ear protection while operating heavy machinery

c. getting lots of sleep

d. turning your radio down

Review (p. 538)

Now that you've finished Section 1, review what you learned by answering the Review questions in your ScienceLog.

_____

Chapter 21, continued

_____

Section 2: Properties of Sound (p. 539)

Name an example of each of the following from your everyday life.

1. a soft sound: _____

2. a loud sound: _____

3. a high-pitched sound: _____

4. a low-pitched sound: _____

The Speed of Sound Depends on the Medium (p. 539)

5. How quickly a sound reaches your ears depends on how loud it is. True or False? (Circle one.)

6. How quickly a sound reaches your ears depends on the medium through which the sound is traveling. True or False? (Circle one.)

7. What did Chuck Yeager accomplish in 1947?

_____

_____

_____

_____

8. In general, as the medium cools, the speed of sound

_____ . (increases or decreases)

9. When particles slow down
   a. they transmit energy more quickly.
   b. they transmit energy more slowly.
   c. they gain kinetic energy.
   d. None of the above

Pitch Depends on Frequency (p. 540)

10. In a guitar, pitch is NOT related to

   a. the frequency of the sound wave.
   b. the thickness of the guitar string.
   c. the number of sound waves produced in a given time.
   d. how far away the guitar is from your ear.

11. Of the animals shown in the graph on page 541:

a. Which animal can hear sounds with the highest pitch?

_____

_____

b. Which animal can hear sounds with the lowest pitch?

_____

_____

12. What frequencies of sound are infrasonic?

_____

_____

_____

_____

13. Name two applications of ultrasonic waves.

_____

_____

_____

_____

_____

14. Look at the Biology Connection on page 541. What is a kidney stone?

a. a bone inside your kidney
b. a kidney-shaped rock used in ultrasound
c. an ancient tool used to harvest kidney beans
d. a calcium-salt deposit in kidneys

15. The Doppler effect does NOT affect how sound is perceived by

a. a pedestrian when a honking driver speeds by.
b. a driver who honks while speeding by a pedestrian.
c. a person sitting in a parked car when a driver honks as he speeds by.
d. a driver speeding past a person who honks the horn of his parked car.

Loudness Is Related to Amplitude (p. 542)

Choose the term in Column B that best matches the definition in Column A. Write your answer in the space provided.

Column A	Column B
_____ **16.** the unit used to express how loud or soft a sound is perceived	**a.** decibel
_____ **17.** how loud or soft a sound is perceived	**b.** loudness
_____ **18.** the maximum distance the particles in a wave vibrate from their rest positions	**c.** amplitude

Use the three diagrams of sounds graphed on the oscilloscopes below to answer questions 19–23.

a. 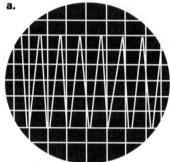 **b.** 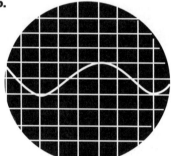 **c.**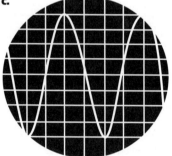

19. Which has the lowest frequency? _____

20. Which has the smallest amplitude? _____

21. Which has the highest frequency? _____

22. Which has the largest amplitude? _____

23. In oscilloscope (b) above, does the left edge of the graph represent a compression, a rarefaction, or neither?

_____

_____

Review (p. 544)

Now that you've finished Section 2, review what you learned by answering the Review questions in your ScienceLog.

Section 3: Interactions of Sound Waves (p. 545)

1. List two reasons why sounds are important to beluga whales.

_____

_____

_____

Reflection of Sound Waves (p. 545)

2. A hard, rigid surface is a better reflector of sound than a soft
   surface is. True or False? (Circle one.)

3. A shout in a gymnasium will usually produce more echoes than
   a shout in an auditorium. True or False? (Circle one.)

4. In Figure 13, how does the Doppler effect help the bat find food?

_____

_____

_____

_____

5. When do you think it is best for ships to use sonar technology
   instead of just relying on eyesight to locate icebergs?

_____

_____

_____

_____

6. Which of the following ways can ultrasonic waves be used in
   medicine? (Circle all that apply.)
   a. performing surgery without making an incision
   b. monitoring the development of an unborn baby
   c. detecting skin cancer
   d. examining internal organs

Review (p. 547)

Now that you've finished the first part of Section 3, review what you
learned by answering the Review questions in your ScienceLog.

Chapter 21, continued

Interference of Sound Waves (p. 548)

7. In _____ interference, two sound waves
   can overlap to produce a softer sound, while in

   _____ interference, two sound waves
   combine to produce a louder sound.

8. A sonic boom is created when an airplane travels faster than the

   speed of _____ . (light or sound)

9. Using Figure 18, explain what happens when a jet flies at super-
   sonic speeds.

   _____

   _____

   _____

   _____

10. In a standing wave, portions of the wave are

    _____ while other portions have a larger
    amplitude due to interference.

11. Suppose you are using a tuning fork to cause a guitar string to
    vibrate without touching it. You get the best results when the
    resonant frequency of the tuning fork matches the

    a. fundamental frequency of the string.
    b. first overtone frequency of the string.
    c. second overtone frequency of the string.
    d. third overtone frequency of the string.

Diffraction of Sound Waves (p. 551)

12. Sound waves with a high frequency have a

    _____ wavelength and

    _____ easily diffracted.
    (long or short, are or are not)

13. When a radio is playing in the next room, which sound waves
    can be heard the best?

    a. sound waves with a high frequency
    b. sound waves with a low frequency
    c. sound waves with a high pitch
    d. sound waves with a small amplitude

_____

Chapter 21, continued

_____

Review (p. 551)

Now that you've finished Section 3, review what you learned by answering the Review questions in your ScienceLog.

Section 4: Sound Quality (p. 552)

1. What is the difference between music and noise?

   a. loudness   **c.** amplitude
   b. pitch   **d.** sound quality

What Is Sound Quality? (p. 552)

2. Why do the same notes sound different on different instruments?

   _____

   _____

   _____

   _____

Sound Quality of Instruments (p. 553)

3. Place each of the following instruments in the correct column:
   drum, guitar, trumpet, cello, bells, tuba, clarinet, banjo, cymbals, violin, saxophone.

String instruments	Wind instruments	Percussion instruments

Fill in each blank in items 4–6 using either *lower pitch* or *higher pitch*.

4. In a string instrument, a longer string has a

   _____ and a thinner string has a

   _____ .

5. In a wind instrument, shortening the air column produces a

   _____ .

6. Among percussion instruments, smaller instruments produce a

   _____ .

Chapter 21, continued

Music or Noise? (p. 555)

7. Which of the following would produce a sound with a random mix of frequencies? (Circle all that apply.)

    a. slamming door
    b. French horn
    c. truck engine
    d. jingling keys

8. What is the difference between the two sound waves shown on the oscilloscopes in Figure 27?

_____

_____

_____

_____

_____

9. When does the amount of noise around you become noise pollution?

    a. when it becomes bothersome
    b. when you can hear it
    c. when it causes health problems
    d. when it is loud

10. According to the Environment Connection on page 555, which of the following groups were NOT affected by the Los Angeles International Airport?

    a. women
    b. men
    c. butterflies
    d. animals

Review (p. 555)

Now that you've finished Section 4, review what you learned by answering the Review questions in your ScienceLog.

CHAPTER

22 DIRECTED READING WORKSHEET

The Nature of Light

As you read Chapter 22, which begins on page 562 of your textbook, answer the following questions.

Strange but True! (p. 562)

1. What causes jaundice?

_____

_____

2. If you were a parent of an infant with jaundice, what method of treatment would you choose? Why?

_____

_____

_____

What Do You Think? (p. 563)

Answer these questions in your ScienceLog now. Then later, you'll have a chance to revise your answers based on what you've learned.

Investigate! (p. 563)

3. Do you think you will be able to tell the difference between fluorescent and incandescent lights using your spectroscope? Why or why not?

_____

_____

Section 1: What Is Light? (p. 564)

4. Name at least two things that produce light.

_____

_____

Light Is an Electromagnetic Wave (p. 564)

5. Sound, unlike light, requires a _____ to travel through.

6. A(n) _____ wave is composed of changing electric and magnetic fields.

Chapter 22, continued

Mark each of the following statements *True* or *False*.

7. _____ A field can exert a push or a pull on an object.

8. _____ Fields are made of matter.

9. _____ Light requires a medium through which to travel.

10. _____ Light waves are considered longitudinal waves because the electric and magnetic fields are perpendicular to each other.

How Light Is Produced (p. 565)

11. An EM wave is produced by the vibration of a(n)

_____ field and a(n)

_____ field.

12. _____ is the emission of energy in the form of EM waves.

Take a moment to look at Figure 2. Then choose the term in Column B that best matches the description in Column A, and write your answer in the space provided. Terms may be used more than once.

Column A	Column B
_____ **13.** move about the nucleus at different distances	**a.** electrons
_____ **14.** tiny "packets" of energy	**b.** photons
_____ **15.** negatively charged particles	
_____ **16.** particles that make up an EM wave	

The Speed of Light (p. 566)

17. Light travels _____ through a vacuum than through a medium such as air, water, or glass. (faster or slower)

18. Why do you see lightning before you hear the thunder that accompanies it?

_____

Review (p. 566)

Now that you've finished Section 1, review what you learned by answering the Review questions in your ScienceLog.

Chapter 22, continued

Section 2: The Electromagnetic Spectrum (p. 567)

1. What does the bee in Figure 4 see that you can't see?

_____

_____

_____

2. Name at least four kinds of EM waves.

_____

_____

Characteristics of EM Waves (p. 567)

3. You can classify an EM wave if you know its

_____ or _____ .

4. The _____ includes the entire range of EM waves.

5. Take a moment to look at the diagram of the electromagnetic spectrum on pages 568 and 569. We can see most of the EM waves in our world. True or False? (Circle one.)

Radio Waves (p. 568)

6. When you listen to AM radio, the music you hear is encoded in the radio waves by the variation of the

_____ of the waves, while FM radio

waves vary in the _____ of the waves.

Mark each of the following statements *AM* or *FM*.

7. _____ Waves can travel a longer distance.

8. _____ Waves have shorter wavelengths.

9. _____ Waves can reflect off the ionosphere.

10. _____ Waves can encode more information.

11. _____ Music sounds better.

12. _____ Waves are used to encode the pictures you see on television.

13. _____ Waves are used to encode the sounds you hear on television.

CHAPTER 22

14. If you watch cable television, how does the television signal get from a distant broadcast studio to your home? List the steps.

_____

_____

_____

_____

_____

_____

Microwaves (p. 570)

Using the information given in Figure 7, fill in the blanks in items 15–18 to complete the description of how a microwave oven works.

15. The microwaves are created by a device called a

_____ , which accelerates charged particles.

16. The waves are reflected into the _____ by

a metal _____ .

17. These waves can travel a distance of several

_____ into the food.

18. They cause molecules of water to _____ .

19. Look at the Brain Food on page 570. Name four household items that use radio waves.

_____

_____

_____

20. What does *radar* stand for?

_____

_____

21. How are microwaves used? (Circle all that apply.)

    a. detecting heat
    b. monitoring airplane movement
    c. calculating the speed of a car
    d. navigating ships at night

Infrared Waves (p. 571)

Mark each of the following statements *True* or *False.*

22. _____ Microwaves have shorter wavelengths than infrared waves do.

23. _____ Infrared radiation can make you feel warm.

24. _____ The warmer something is, the more infrared radiation it gives off.

25. _____ With the right equipment, you can "see" in the dark using infrared radiation.

Review (p. 571)

Now that you've finished the first part of Section 2, review what you learned by answering the Review questions in your ScienceLog.

Visible Light (p. 572)

26. The shortest wavelengths of visible light, which carry the

_____ energy, are seen as the color

_____ .

27. The longest wavelengths of visible light are seen as the color

_____ .

28. Look at the Biology Connection on page 572. What kind of EM wave fuels photosynthesis?

    a. microwaves   **c.** visible light

    b. infrared light   **d.** ultraviolet light

29. Who is Roy G. Biv, and why should you remember his name?

_____

_____

_____

_____

Ultraviolet Light (p. 573)

30. Name two benefits of ultraviolet light.

_____

_____

_____

31. Name two hazards of ultraviolet light.

_____

_____

X Rays and Gamma Rays (p. 574)
32. How are X rays useful in the medical field?

_____

_____

33. Why do you often need to wear a lead apron when getting an
X ray?

_____

_____

34. Which of the following is NOT true of gamma rays?
   a. They are used to treat cancer.
   b. They carry the least amount of energy of the EM waves.
   c. They penetrate materials easily.
   d. You are exposed to them every day.

Review (p. 574)
Now that you've finished Section 2, review what you learned by
answering the Review questions in your ScienceLog.

Section 3: Interactions of Light Waves (p. 575)
1. How is the special layer of cells in the back of a cat's eyes useful?

_____

_____

Reflection (p. 575)
2. When you look in a mirror you see light that has been reflected

twice. True or False? (Circle one.)

3. What is the law of reflection?

_____

_____

_____

_____

Chapter 22, continued

_____

4. In Figure 14, the _____ is the line perpendicular to the mirror's surface.

5. What kind of surface allows you to see your reflection?

_____

_____

Absorption and Scattering (p. 576)

Decide if each of the following refer to scattering or absorption. Write *S* for scattering and *A* for absorption.

6. _____ EM waves transfer energy to particles in matter.

7. _____ Particles of matter that have absorbed energy release light energy.

8. _____ On a dark night, you can see objects outside of a flashlight beam.

9. _____ The sky appears blue.

10. _____ Air particles absorb the energy from light, making it less bright.

Refraction (p. 577)

11. Refraction is caused by a variation in the

_____ of light as it passes from one medium to another.

12. The speed of light traveling through glass is slower than the speed of light traveling through air. True or False? (Circle one.)

13. Optical illusions occur because your brain interprets light as

traveling in _____ lines.

14. Color separation occurs when light is refracted because light with a long wavelength is bent more than light with a short wavelength. True or False? (Circle one.)

Diffraction (p. 579)

15. The _____ of waves around barriers and through openings is called diffraction.

16. In order for the greatest amount of diffraction to occur, an

opening must be the same size or _____ than the wavelength of the light passing through it.

---

17. A greater amount of diffraction occurs when light passes through a narrower opening. True or False? (Circle one.)

Interference (p. 579)

18. The wave that results from _____ interference has a greater amplitude than the individual waves that combined to form it.

19. The result of _____ interference is dimmer light.

20. Why do you not see constructive and destructive interference of white light?

_____

_____

_____

Use the information in Section 3 to answer the following questions. Choose the word in Column B that best matches the definition or example in Column A, and write your answer in the space provided.

Column A	Column B
_____ **21.** transfer of energy from light waves to particles of matter	**a.** interference
_____ **22.** bending of light passing through an opening	**b.** scattering
_____ **23.** bending of light passing into a different material	**c.** absorption
_____ **24.** a wave bouncing off an object	**d.** diffraction
_____ **25.** waves overlapping and combining	**e.** refraction
_____ **26.** the release of light energy by particles of matter that have absorbed energy	**f.** reflection

Review (p. 580)

Now that you've finished Section 3, review what you learned by answering the Review questions in your ScienceLog.

Section 4: Light and Color (p. 581)

1. What is white light made of?

_____

_____

Light and Matter (p. 581)

2. What are the three ways light can interact with matter?

_____

3. The three interactions you listed in question 2 can all occur at the same time. True or False? (Circle one.)

In items 4–7, complete each sentence using the word *transparent, translucent,* or *opaque.*

4. Most windows are _____ .

5. A frosted glass window is _____ .

6. A wooden door is _____ .

7. An object that is _____ transmits light but scatters the light as it passes through.

Colors of Objects (p. 582)

8. An object appears a certain color as a result of

    a. the wavelengths of light that reach your eyes.
    b. the wavelengths of light that reach the object.
    c. the size of the object.
    d. the size of the light source.

9. The color of an opaque object is the color of the light that is

_____ by the object. The object

_____ any remaining colors.

10. What colors of light are reflected by an opaque white object?

_____

11. What colors of light are reflected by an opaque black object?

_____

12. The color of a transparent object is the color of the light that is

_____ through the object. The object

_____ any remaining colors.

Mixing Colors of Light (p. 584)

13. The three primary colors of light are

_____ , _____ ,

and _____ .

14. Why are the colors you wrote in item 13 called the primary colors of light?

_____

_____

15. What happens when you add two primary colors of light together?

_____

_____

16. The three secondary colors of light are

_____ , _____ ,

and _____ .

Mixing Colors of Pigment (p. 584)

17. How does a pigment give a substance color?

_____

_____

18. Why is the mixing of colors of paint called color subtraction?

_____

_____

_____

_____

_____

19. What are the three primary pigments?

_____

20. Take a moment to read the Geology Connection on page 585. What does azurite change into over time?

_____

Review (p. 585)

Now that you've finished Section 4, review what you learned by answering the Review questions in your ScienceLog.

DIRECTED READING WORKSHEET

Light and Our World

As you read Chapter 23, which begins on page 592 of your textbook, answer the following questions.

Imagine . . . (p. 592)

1. Using the retroreflector scientists have learned that the moon is

moving _____ the Earth.

What Do You Think? (p. 593)

Answer these questions in your ScienceLog now. Then later, you'll have a chance to revise your answers based on what you've learned.

Investigate! (p. 593)

2. What is the purpose of this activity?

_____

_____

_____

_____

Section 1: Light Sources (p. 594)

3. Visible light

    a. covers the entire electromagnetic spectrum.
    b. covers most of the electromagnetic spectrum.
    c. covers a small part of the electromagnetic spectrum.
    d. is not a part of the electromagnetic spectrum.

Light Source or Reflection? (p. 594)

4. Look at the Astronomy Connection on page 594. The sun and the moon are two celestial objects visible from Earth. Which one is luminous, and which one is illuminated?

_____

_____

_____

For each of the following, write *Yes* if it is a light source or *No* if it is not a light source in the space provided.

5. _____ light bulb

6. _____ wooden bookshelf

7. _____ burning candle

8. _____ glow-in-the-dark rubber ball

9. _____ coin

10. _____ firefly

11. _____ mirror

Producing Light (p. 595)

12. What does an incandescent light emit? (Circle all that apply.)

    a. electricity
    b. thermal energy
    c. light
    d. halogens

13. The inside of a fluorescent light tube is coated with phosphor. What would happen if the phosphor were removed?

_____

_____

_____

_____

_____

_____

In a neon light, the type of gas in the tube determines the color of the light produced. Match the gas in Column A with the color of light in Column B, and write your answer in the space provided.

Column A	Column B
____ **14.** neon	**a.** blue
____ **15.** sodium	**b.** red
____ **16.** argon and mercury	**c.** yellow

---

17. What types of vapor do street lights usually contain?

_____

_____

Review (p. 597)

Now that you've finished Section 1, review what you learned by answering the Review questions in your ScienceLog.

Section 2: Mirrors and Lenses (p. 598)

1. A rearview mirror reverses an image left to right. How would the word *LAMB* appear if you viewed it in a rearview mirror? (Hint: look at the figure of the ambulance on the right margin.)

_____

Rays Show the Path of Light Waves (p. 598)

2. The path a light wave takes is called a _____

because it travels along a _____ path.
(ray or arc, straight or curved)

Mirrors Reflect Light (p. 599)

3. Imagine that Sarah, a new college student, goes to the bookstore and buys herself a sweatshirt with the name of her college on it. Back in her dorm room, she puts on the shirt and looks in the mirror (a plane mirror). Describe the image she sees.

_____

_____

_____

Concave mirrors produce different types of images depending on where the reflected object is placed. Match the object placement in Column A with the image created in Column B, and write your answer in the space provided.

Column A	Column B
____ **4.** in front of the focal point	**a.** real image
____ **5.** at the focal point	**b.** virtual image
____ **6.** beyond the focal point	**c.** no image

7. Imagine you are at a carnival funhouse looking at a mirror. No matter how far you stand away from the mirror, your image is upright and small. What type of mirror are you looking at?

_____

_____

_____

Review (p. 602)
Now that you've finished the first part of Section 2, review what you learned by answering the Review questions in your ScienceLog.

Lenses Refract Light (p. 603)

8. How are lenses classified?

    a. by their color
    b. by the type of light they refract
    c. by their size
    d. by their shape

9. If you shined two laser pointers in parallel at a convex lens,

the lens would _____ the light and

the beams would cross at the _____ .
(reflect or refract, optical axis or focal point)

10. Why can't you form a real image with a concave lens?

_____

_____

_____

_____

Review (p. 604)
Now that you've finished Section 2, review what you learned by answering the Review questions in your ScienceLog.

Section 3: Light and Sight (p. 605)

How You Detect Light (p. 605)

1. What controls the size of the pupil?

_____

2. In bright light the muscles of the iris contract, making the pupil

_____ . (larger or smaller)

3. In low light the muscles of the iris relax, making the pupil

_____ . (larger or smaller)

4. What do the muscles surrounding the lens do?

_____

_____

_____

Common Vision Problems (p. 606)

Imagine that two of your friends, Emily and Harold, left their glasses at your house. Emily is nearsighted. Harold is farsighted. One of the pairs of glasses has convex lenses, the other has concave. Match the description in Column A with the friend in Column B, and write your answer in the space provided. Answers can be used more than once.

Column A	Column B
_____ **5.** wears convex lenses	**a.** Emily
_____ **6.** wears concave lenses	**b.** Harold
_____ **7.** eyes are too long	
_____ **8.** eyes are too short	

9. What specific problem would a driver who is red-green colorblind face?

   a. reading stop signs
   b. seeing double
   c. distinguishing between stop and go lights
   d. checking his or her blind spot

CHAPTER 23

10. Explain why there are not corrective lenses for colorblindness as there are for nearsightedness and farsightedness.

_____

_____

_____

Review (p. 607)

Now that you've finished Section 3, review what you learned by answering the Review questions in your ScienceLog.

Section 4: Light Technology (p. 608)

Optical Instruments (p. 608)

The camera is an optical instrument that is similar in some ways to your eye. Match the camera parts in Column A with the corresponding eye structure in Column B, and write your answer in the space provided.

Column A	Column B
____ **1.** aperture	**a.** retina
____ **2.** shutter	**b.** pupil
____ **3.** lens	**c.** cornea and lens
____ **4.** film	**d.** no corresponding structure

5. An astronomer is looking at light from a distant star through a refracting telescope. Below is a list of items the light ray would encounter between the star and the astronomer's retina. In which order would the light ray strike or pass through them? Write the items in order in the spaces provided, and indicate if the item would refract or reflect the light ray.

telescope objective lens
cornea and lens of astronomer's eye
telescope eyepiece lens

_____

_____

_____

6. Another astronomer is looking at light from the same distant star through a reflecting telescope. Below is a list of items the light ray would encounter between the star and the astronomer's retina. Write the items in order in the spaces provided, and indicate if the item would refract or reflect the light ray.

cornea and lens of the astronomer's eye
telescope plane mirror
telescope eyepiece lens
telescope concave mirror

_____

_____

_____

_____

7. Light microscopes are similar in construction to

_____ .

(refracting telescopes or reflecting telescopes)

Lasers and Laser Light (p. 609)

8. Unlike white light, laser light is _____,
meaning that all the light waves move together away from the source.

9. Using Figure 23, put the following items in the correct order to show how a laser is created.

_____ Excited neon atoms release photons, which strike other atoms.

_____ Some photons leak out of a partially coated mirror to create a beam.

_____ Plane mirrors on the ends of the laser reflect photons back and forth.

_____ An electric current excites neon atoms in a gas-filled tube.

10. What causes the laser light in Figure 23 to become brighter?

_____

_____

_____

_____

11. What would happen to the hologram being made in Figure 24 if the beam splitter were removed?

   a. This would have no effect.
   b. The laser would quit shining due to the interference pattern.
   c. There would be no interference pattern formed on the film.
   d. The object would melt.

12. What common household device contains a laser?

   _____

   _____

Fiber Optics (p. 612)

13. Why have fiber-optic cables largely replaced copper telephone cables?

   _____

   _____

   _____

14. Light cannot escape an optical fiber because of total

   _____ .

Polarized Light (p. 612)

15. Polarizing sunglasses decrease _____ from horizontal surfaces by reflecting light that vibrates

   _____ .

16. The two photos in Figure 28 were taken by the same camera from the same angle. Why are they different?

   _____

   _____

   _____

Review (p. 613)

Now that you've finished Section 4, review what you learned by answering the Review questions in your ScienceLog.